ACCLAIM FOR *THE COSMIC VIEW OF ALBERT EINSTEIN*

Periodically throughout human history individuals have appeared whose insight and imagination change our understanding in a way that can only be called revolutionary. Einstein revolutionized our scientific understanding of space and time with his use of "thought experiments", which particularly intrigue me. These impressive images of the cosmos in which we participate are indeed a fitting accompaniment to this collection of quotations from his work.

—His Holiness the 14th Dalai Lama Tenzin Gyatso
1989 Nobel Peace Laureate

The Cosmic View of Albert Einstein *is one of the most inspiring books that I have ever seen. The photographs of the stellar constellations, the galaxies and the individual stars literally give me goose pimples. I myself cannot understand the concept of infinity, a notion that has plagued me all my life, and if contemplated too deeply would literally drive me out of human comprehension. The quotes of Albert Einstein interdigitated with the absolute wonders of the universe make the tome a precious and absolutely unique commodity. What a curious and wonderful man was he, full of the compassion of human kindness and sagacity. He is in a way a modern day prophet of the likes of Buddha, Jesus and Gandhi. He knew much but at the same time he recognized that he really knew and understood little about the cosmic puzzle as he stood in wonder before it.*

—Helen Caldicott M.D.
author, co-founder Physicians for Social Responsibility, named by the Smithsonian Institute as one of the most influential women of the 20th century

The thoughts and experiences of Albert Einstein on religion, existence and morality, as revealed in this magnificent volume, are those of a deeply insightful, moral and intelligent man of the cosmos. It is a vital read for all who wrestle with the greatest questions and dilemmas of our troubled civilization.

—Edgar Mitchell, D.Sc.
Apollo 14 astronaut, 6th man on the Moon, founder Noetic Institute, author

THE

COSMIC

VIEW

OF

ALBERT

EINSTEIN

edited by **WALT MARTIN and MAGDA OTT**

STERLING
New York

STERLING
New York

An Imprint of Sterling Publishing
387 Park Avenue South
New York, NY 10016

Compilation © 2013 by Walter Martin and Magda Ott
Credits for text on page 137 and photos on page 149

ISBN 978-1-4549-0776-3

Distributed in Canada by Sterling Publishing
c/o Canadian Manda Group, 165 Dufferin Street
Toronto, Ontario, Canada M6K 3H6
Distributed in the United Kingdom by GMC Distribution Services
Castle Place, 166 High Street, Lewes, East Sussex, England BN7 1XU
Distributed in Australia by Capricorn Link (Australia) Pty. Ltd.
P.O. Box 704, Windsor, NSW 2756, Australia

Design and production: Susan Welt for gonzalez defino, ny / gonzalezdefino.com

For information about custom editions, special sales, and premium and
corporate purchases, please contact Sterling Special Sales at 800–805–5489
or specialsales@sterlingpublishing.com

Manufactured in China

2 4 6 8 10 9 7 5 3 1

www.sterlingpublishing.com

In memory of Dr. James Van Allen,
who never lost a holy curiosity

Contents

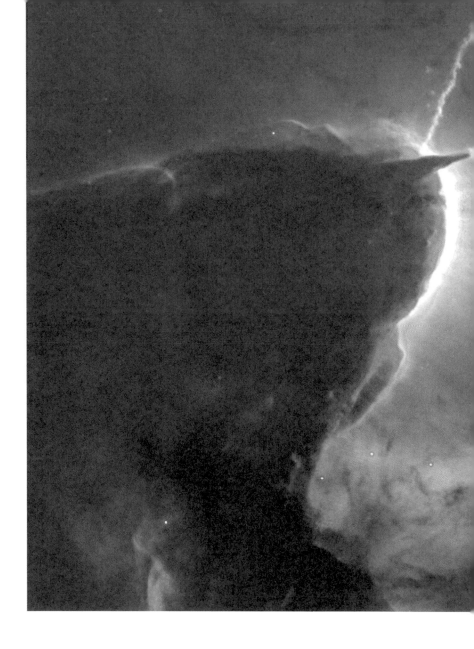

▲ **THE TRIFID NEBULA** Resembling a whimsical character
out of Alice in Wonderland, this Hubble Space Telescope image
actually reveals a stellar nursery being torn apart by radiation
from a nearby, massive star. Located about 9,000 light-years from
Earth, the Trifid resides in the constellation Sagittarius.

This is a beautiful book. But it is far more than that. It presents a wealth of splendid images of the wonders of nature, in their infinite complexity and enormous dimensions. These images challenge the reader to share in the fruits of modern astronomy and to grapple with their humanistic significance. The editors' perceptive quotations from the writings of Albert Einstein respond to that challenge. Einstein's thoughts guide the reader to a profound appreciation of our universe and to belief in a Divine and Omniscient Power.

James A. Van Allen
discoverer of the radiation belt around
the Earth (called the Van Allen belt)
that creates the polar auroras

Foreword

Einstein's Cosmic View

After Einstein's death in 1955, cartoonist Herblock, in the April 19 issue of the Washington Post, published one of his most famous cartoons. In it, he depicted our little planet floating among other spherical objects in outer space, with a large square note tacked onto Earth: "Albert Einstein Lived Here." While simple, this intriguing cartoon communicated the complex nature of the century's most famous physicist: his importance to the world as a human being, his contributions to understanding fundamental truths about our universe, and his view of himself as a citizen of the whole world, not only one nation.

In an appropriate juxtaposition of wisdom, intellect, technology, and art, the editors' compilation of Einstein's most memorable words and photographs by NASA, other observatories around the world, and amateur astronomers vividly captures the beauties of our expanding and dynamic Universe. "The eternal mystery of the world is its comprehensibility. The fact that it is comprehensible is a miracle," Einstein mused in 1936. These photos and the work of the scientists and technical experts behind them—artists all—are proof of humankind's desire to comprehend the miraculously changing canvas we call our cosmos.

Albert Einstein, the supernova among physicists, is best known for his so-called genius, pacifism, and, in his later years, humanitarian and political activism. Though his achievements are manifold, enough to make the most accomplished among us blush, he was in fact a modest and humble human being, making his way through life like the rest of us, often bumbling and making mistakes along the way. He was, however, wise enough to change his mind as circumstances and the passage of time dictated, both in his physics and in his worldview.

As a prelude to the photographs that follow, it seems appropriate to expound briefly on Einstein's humanitarian proclivities. His convictions and beliefs might best be summarized in his "cosmic view" about religion, pacifism, the social responsibility of scientists, and in his passionate desire for a politically neutral world government established to protect the planet and us all from the baser in-stincts of humankind.

Einstein's "Cosmic Religion"

The dominant effect of the photos in this book is to inspire wonder and awe, words Einstein used in his attempt to define his faith in the power and laws of Nature. This he called his "cosmic religion." The subject of Einstein's religion is a thorny one, and we can only judge his spiritual feelings by reading about them in the words he left behind, many of which accompany the photographs in this book. As someone who is fairly familiar with Einstein's writings and personality, I venture to say that by his profession of a "cosmic" religion, Einstein most likely meant to convey that it is possible to be religious—that is, not an atheist—without belief in the "personal" God that most societies throughout the world see as the "real" God.

To Einstein's mind, he was not an atheist: "Even in view of the harmony in the cosmos which I, with my limited human mind, am able to recognize, there are still people who say there is no God. But what makes me really angry is that they quote me for support of such views." "Then there are the fanatical atheists whose intolerance is the same as that of the religious fanatics, and it springs from the same source…. They are creatures who can't hear the music of the spheres." The "music of the spheres" idea dates back at least to the sixteenth century—the music of the spheres is said to order the heavens, and music tempers human passions and social forces. With words such as these, written around 1941, Einstein rejected atheism. At best, however, he considered himself an agnostic: "My position concerning God is that of an agnostic. I am convinced that a vivid consciousness of the primary importance of moral principles for the betterment and ennoblement of life does not need the idea of a law-giver, especially a law-giver who works on the basis of reward and punishment," he wrote in 1950.

Einstein's idea of religion, rather than fashioned by dogma dictated, prescribed, and refashioned over the ages by millions of self-appointed experts and unquestioning believers, is based on a more constant theme—that of nature and her almost unwavering, harmonious laws. Notions of nature as a sort of God have been around for centuries, the most prominent proponent being the Jewish-Dutch philosopher Baruch Spinoza, who greatly influenced Einstein's religious thinking. "I believe in Spinoza's God, who reveals himself in the lawful harmony of the world, not in a God who concerns himself with the fate and the doings of mankind." In this way, Einstein was unifying science and religion, and referred to himself as a "deeply religious nonbeliever." Moreover, being open-minded and inclusive in his worldview, he found Jesus, Buddha, and Moses equally compelling as prophets.

Einstein was in wonder and awe that "the Old One," as he referred to his God, had set an almost perfect system of order in motion since the earliest times of the big bang. This system has persevered through eons of physical changes, and, in the case of Earth at least, through biological transformations and evolution. Through these immutable laws of nature, the universe has been able to survive to the present day. In more recent times, humankind, often through the exploitation of its natural resources, has been able to tamper with natural laws in the name of progress, often resulting in benefit to people but in harm to the planet. In today's world, Einstein would surely speak out for a balance that, through some sacrifice on the part of overly zealous consumers in some parts of the world, is surely possible.

Pacifism, Social Responsibility of the Scientist, and World Government

Einstein was a lifelong pacifist except during the World War II era, when Adolf Hitler forced him to compromise his long-held beliefs. "My pacifism is an instinctive feeling, a feeling that possesses me because the murder of people is disgusting," he wrote in 1929. "My attitude is not derived from any intellectual theory but is based on my deepest antipathy to every kind of cruelty and hatred." He considered himself a "militant pacifist," one willing to fight for peace, and he admired the peaceful-resistance teachings of Mahatma Gandhi, who said that "nonviolence is the greatest force at the disposal

▶ **A FIREFLY STUDDED MEADOW** The wondrous glow of Lampyridae (loosely translated as "shining fire") beetles, a bioluminescent insect that mimics the stars, is 95–99% efficient with about 1–5% heat loss; even the most advanced technologies have not been able to reproduce that. Photograph taken in the firefly's natural environment, without flash and without digital manipulation by R. Schreiber in Iowa.

of mankind." Had he lived into the era of Martin Luther King, Jr., he surely would have supported King's peaceful tactics and agenda as well. Einstein also often spoke of the responsibility of scientists and policy makers to make the best use of new discoveries, for peaceful purposes rather than war, and for the benefit of all humankind. In August 1948, three years after the end of World War II and in an uncertain new atomic age, he released a message to fellow intellectuals: "We scientists, whose tragic destination has been to help in making the methods of annihilation more gruesome and more effective, must consider it our solemn and transcendent duty to do all in our power in preventing these weapons from being used for the brutal purpose for which they were invented. What task could possibly be more important for us?"

Einstein felt great remorse about the contribution of physics that led to the bomb, and spent the last ten years of his life fighting for the peaceful uses of atomic energy and a "supranational" organization—that is, a kind of world government whose purpose would be the control of weapons and the guarantee of individual freedoms. Such an international body, he felt, was the only possible vehicle toward world peace. He appended his last signature to a nonscientific statement that came to be called the Russell-Einstein Manifesto, one of the most important documents of the twentieth century which remains highly relevant in the twenty-first century. It was formally issued three months after his death by philosopher and peace activist Bertrand Russell. This document was a call to all nations "to renounce nuclear weapons as part of a general reduction of armaments" and was signed by nine other prominent scientists. They asked scientists of the world, as well as the general public, to subscribe to the following resolution: "In view of the fact that in any future world war nuclear weapons will certainly be employed, and that such weapons threaten the continued existence of mankind, we urge the governments of the world to realize, and to acknowledge publicly, that their purpose cannot be furthered by a world war, and we urge them, consequently, to find peaceful means for the settlement of all matters of dispute between them." However, the governments of the world paid no attention.

Today, Einstein continues to be honored for his unwavering if unsuccessful humanitarian struggles to achieve peace, world order, and international cooperation, and for his passionate opposition to McCarthyism, racial segregation, ethnic discrimination, and his support of human rights throughout the world. As readers peruse or inspect the spectacular photographic creations that follow, they are certain to be filled with the awe and wonder that Einstein felt when he contemplated nature, no matter if their own religious beliefs are different from his. As we see that we are but a tiny note in the music of the spheres, all Earthlings should redouble their efforts to come together as one people on Earth, here to protect, preserve, and revere our physical space as well as our fellow creatures.

Alice Calaprice, *author of*
The Ultimate Quotable Einstein, *is a former*
senior editor at Princeton University Press in charge
of The Collected Papers of Albert Einstein.

"I hold that mankind is approaching an era
in which peace treaties will not only be recorded
on paper, but will also become inscribed in
the hearts of men." *(1946)*

Albert Einstein's Legacy

This book presents Albert Einstein's "cosmic view" through direct quotations from many different sources (essays, books, letters, and interviews,) and illustrates them with spectacular astronomical images. The book is organized in such a way as to give the reader insight into Einstein's beliefs, as well as a sense of the man, and thereby arouse the reader's curiosity to explore deeper into the richness of his contributions to modern science and modern thought.

Einstein was not only a great scientist but also an artist in his use of words. His writings reveal a deeply thoughtful man, who did not believe in a personal God but that "a spirit is manifest in the laws of the Universe—a spirit vastly superior to that of man, and one in the face of which we with our modest powers must feel humble."

His fame rests on his two published theories and the astronomical observations of physical phenomena which confirm them. He overthrew a view of the universe that had endured for three centuries. In its place he constructed a new one: a profoundly strange and beautiful universe, where time is another dimension and where there is no standard of reference. A meter stick is only a meter while it is at rest. Move it and it becomes shorter, and the faster it is moved the shorter it becomes.

Time is also relative. In this Alice In Wonderland-like universe, time is no longer an unalterable absolute measure. In motion, every body has its own time which elapses more slowly as the body moves more rapidly. The ultimate barrier is the speed of light. No material object can go as fast; for as speed increases so does mass until—at the speed of light—mass becomes infinite and time stands still.

"Curiouser and curiouser!" as Alice would exclaim, space-time itself is warped by matter in Einstein's universe. This warping or curving of space-time is gravity, and can become so powerful that it crushes matter into a black hole, out of which nothing, not even light, can escape.

The year 1905 is called his "miraculous year" because Einstein published the *Special Theory of Relativity* (which laid the groundwork for the *General Theory of Relativity*) and scientific papers on Brownian motion and the photoelectric effect. Special and general relativity are essential to modern cosmology—the study of the origin, structure, and space-time relationships of the universe—revealing that matter is congealed energy ($E=MC^2$), the existence of gravitational waves, black holes, and the most compelling theory of creation, the big bang. His explanation of the photoelectric effect helped others develop quantum mechanics: the physics of atomic structure and propagation of energy.

Einstein stubbornly rejected quantum theory because of its inherent indeterminacy, stating that "God does not play dice." Yet he was wrong. The universe is fueled by indeterminacy (uncertainty principle, entanglement, virtual particles, vacuum fluctuations, etc.), pure creativity that springs from within itself by itself through itself, expressing that infinite creativity through the laws of nature. Nevertheless, Einstein deeply appreciated natural law and "the illimitable superior spirit" it reveals. He spent his entire life seeking to understand those laws in hope of glimpsing God's thoughts. "I want to know how God created this world. I am not interested in this or that phenomenon, in the spectrum of this or that element. I want to know His thoughts, the rest are details."

Useful gadgets that ingenious inventors can make from Einstein's theories seem limitless: solar-powered devices, GPS units, digital cameras, lasers in DVD players. A new generation of relativistic computer chips could soon result in processors that run much faster than current models. These would radiate far less heat and consume far less power. Other Einstein-influenced inventions might emerge from research labs in the future. Nanotechnologists are planning devices that could speed up DNA analysis by harnessing the random motion of molecules the result of a phenomenon first correctly explained by Einstein in 1905. Around the world, laboratories are creating exotic forms of matter envisioned by Einstein in 1925 in one of his classic "thought experiments." Coherent ultracold atoms—matter's equivalent to laser beams—could find use in portable atomic clocks, superprecise gyroscopes for navigation, and gravity sensors for mapping mineral lodes and oil fields.

Einstein's greatest scientific legacy may be the unified field, which would weave all laws of nature into a single coherent theory. His solitary effort during the last thirty years of his life has inspired others in the chase after the theory of everything. Before Einstein's death, little was known about the nuclear force; part of the puzzle had therefore been missing. String theory is now the leading candidate for unification. Relativity and quantum mechanics, the two great pillars of modern science, may eventually be integrated into a single unified field theory. It would summarize all fundamental physical knowledge and rank as the greatest achievement of science, allowing us, as Einstein believed, "to read the mind of God." Today, a hundred years later, he is still helping us understand the universe's ever-expanding wonders.

But as impressive as Einstein's discoveries of physical phenomena of the cosmos are, his often overlooked lifelong passion for human rights and world peace is of even greater importance. Not to know this does a disservice to him and everyone who imagines Einstein as a rumpled genius, too distracted to notice his surroundings, or even worse, "the father of the bomb."

Nobel Peace Prize laureate Joseph Rotblat, one of the *Russell-Einstein Manifesto* cosigners, wrote about Einstein in 2005: "He was quite the

▲ **ALBERT EINSTEIN RECEIVING TELESCOPE AS GIFT.** Dr. Albert Einstein peers into the eyepiece of an eight-inch Newtonian F-8 telescope, handmade by Mr. Zvi Gezari, left, as Mrs. Sharp, right watches. Dr. Einstein accepted the telescope on behalf of the Elsa and Albert Einstein School in Ben Shemen, Israel. The school provides scientific training for children of 21 different nationalities.

opposite of what people think about scientists... I admire him not only as a great man of science but also as a great human being."

Albert Einstein made his views clear about science's responsibility to humanity. At a visit to the California Institute of Technology in Pasadena, he told Caltech students: "Why does applied science bring us so little happiness? The simple answer is that we have not yet learned to make proper use of it. In time of war it has given men the means to poison and mutilate one another. In time of peace it has made our lives hurried and uncertain. It has enslaved us to machines. The chief objective of all technological effort must be concern for mankind. Never forget this when you are pondering over your diagrams and equations."

Walt Martin, Magda Ott

editors

We shall not cease from exploration

And the end of all our exploring

Will be to arrive where we started

And know the place for the first time.

—*T. S. Eliot, "Little Gidding V,"* Four Quartets *(1943)*

Part I

▲ **SPIRAL GALAXY NGC 1232** is one of the most inspiring sights
in the universe. As its two principal arms unwound they broke into
many segments that now dapple the whole body of the galaxy—giving
it an unusually complex structure. Myriad blue knots punctuate the
arms at the birthing sites of stars.

Cosmic Religion

My religiosity consists of a humble admiration of the infinitely superior spirit that reveals itself in the slight details we are able to perceive of the knowable world with our frail and feeble minds. That deeply emotional conviction of the presence of a superior reasoning power, which is revealed in the incomprehensible universe, forms my idea of God.

. . .

The most beautiful and most profound emotion we can experience is the sensation of the mystical. It is the sower of all true science. He to whom this emotion is a stranger, who can no longer wonder and stand rapt in awe, is as good as dead. To know that what is impenetrable to us really exists, manifesting itself as the highest wisdom and the most radiant beauty which our dull faculties can comprehend only in their most primitive forms—this knowledge, this feeling is at the center of true religiousness.

. . .

A human being is a part of the whole, called by us "Universe," a part limited in time and space. He experiences himself, his thoughts and feelings as something separated from the rest—a kind of optical delusion of his consciousness. This delusion is a kind of prison for us, restricting us to our personal desires and to affection for a few persons nearest to us. Our task must be to free ourselves from this prison by widening our circle of compassion to embrace all living creatures and the whole of nature in its beauty.

Beginnings

When I was a fairly precocious young man I became thoroughly impressed with the futility of the hopes and strivings that chase most men restlessly through life. Moreover, I soon discovered the cruelty of that chase, which in those years was much more carefully covered up by hypocrisy and glittering words than is the case today. By the mere existence of his stomach everyone was condemned to participate in that chase. The stomach might well be satisfied by such participation, but not man insofar as he is a thinking and feeling being. As the first way out there was religion, which is implanted into every child by way of the traditional education-machine. Thus I came —though the child of entirely irreligious (Jewish) parents—to a deep religiousness, which, however, reached an abrupt end at the age of twelve. Through the reading of popular scientific books I soon reached the conviction that much in the stories of the Bible could not be true. The consequence was a positively fanatic [orgy of] freethinking coupled with the impression that youth is intentionally being deceived by the state through lies; it was a crushing impression. Mistrust of every kind of authority grew out of this experience, a skeptical attitude toward

◄ **IN THIS STUNNING PICTURE** of the giant galactic nebula NGC 3603, NASA's Hubble space telescope captures various stages of the life cycle of stars. Dark clouds at the upper right are so-called Bok globules, probably in an early stage of star formation. The star's grayish-blue circumstellar ring of glowing gas and bipolar outflows (blobs to the upper right and lower left) indicate the chemically enriched material made in a supernova explosion, marking the end of the life cycle.

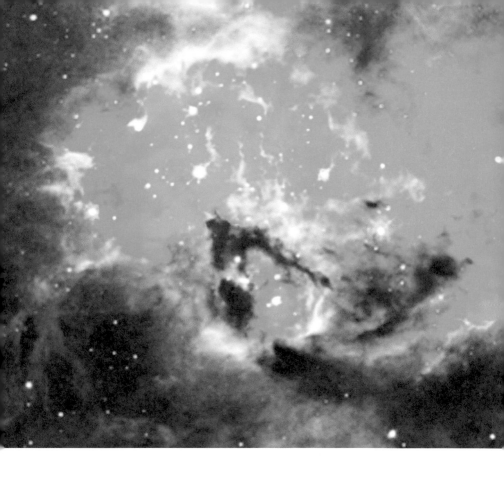

the convictions that were alive in any specific social environment—an attitude that has never again left me, even though, later on, it has been tempered by a better insight into the causal connections.

It is quite clear to me that the religious paradise of youth, which was thus lost, was a first attempt to free myself from the chains of the "merely personal," from an existence dominated by wishes, hopes, and primitive feelings. Out yonder there was this huge world, which exists independently of us human beings and which stands before us like a great, eternal riddle, at least partially accessible to our inspection and thinking. The contemplation of this world beckoned as a liberation, and I soon noticed that many a man whom I had learned to esteem and to admire had found inner freedom and security in its pursuit. The mental grasp of this extra-personal world within the frame

▲ **NGC 2074 IN THE LARGE MAGELLANIC CLOUD** In commemoration of NASA's Hubble Space Telescope completing its 100,000th orbit in its eighteenth year of exploration and discovery, scientists have aimed Hubble to take a snapshot of a dazzling region of celestial birth and renewal. The region is a firestorm of raw stellar creation, perhaps triggered by a nearby supernova explosion.

of our capabilities presented itself to my mind, half consciously, half unconsciously, as a supreme goal. Similarly motivated men of the present and of the past, as well as the insights they had achieved, were the friends who could not be lost. The road to this paradise was not as comfortable and alluring as the road to the religious paradise; but it has shown itself reliable, and I have never regretted having chosen it.

The Meaning of Life

What is the meaning of human life, or of organic life altogether? To answer this question at all implies a religion. Is there any sense then, you ask, in putting it? I answer, the man who regards his own life and that of his fellow creatures as meaningless is not merely unfortunate but almost disqualified for life.

·　　·　　·

The life of the individual has meaning only insofar as it aids in making the life of every living thing nobler and more beautiful. Life is sacred, that is to say, it is the supreme value, to which all other values are subordinate.

·　　·　　·

I feel myself so much a part of everything living that I am not in the least concerned with the beginning or ending of the concrete existence of any one person in this eternal flow.

·　　·　　·

There are moments when one feels free from one's own identification with human limitations and inadequacies. At such moments one imagines that one stands on some small spot of a small planet gazing in amazement at the cold yet profoundly moving beauty of the eternal, the unfathomable. Life and death flow into one and there is neither evolution nor destiny; only being.

◄ **EARTH AND ITS MOON FROM GALILEO SPACECRAFT.**
Eight days after its encounter with the Earth, the Galileo
spacecraft was able to look back and capture this remarkable
view of the Moon in orbit around the Earth, taken from a
distance of about 6.2 million kilometers (3.9 million miles).

Self-Portrait

Of what is significant in one's own existence one is hardly aware, and it certainly should not bother the other fellow. What does a fish know about the water in which he swims all his life?

The bitter and the sweet come from the outside, the hard from within, from one's own efforts. For the most part I do the thing which my own nature drives me to do. It is embarrassing to earn so much respect and love for it. Arrows of hate have been shot at me too; but they never hit me, because somehow they belonged to another world, with which I have no connection whatsoever.

I lived in that solitude which is painful in youth, but delicious in the years of maturity.

◄ **ROSETTE NEBULA, NGC 2237 is a large emission nebula located 3000 light-years away. The great abundance of hydrogen gas gives NGC 2237 its red color in most photographs. The wind from the open cluster of stars known as NGC 2244 has cleared a hole in the nebula's center. The filaments of dark dust lanes run through the nebula's gases. The origin of recently observed fast-moving molecular knots in the Rosette remains under investigation.**

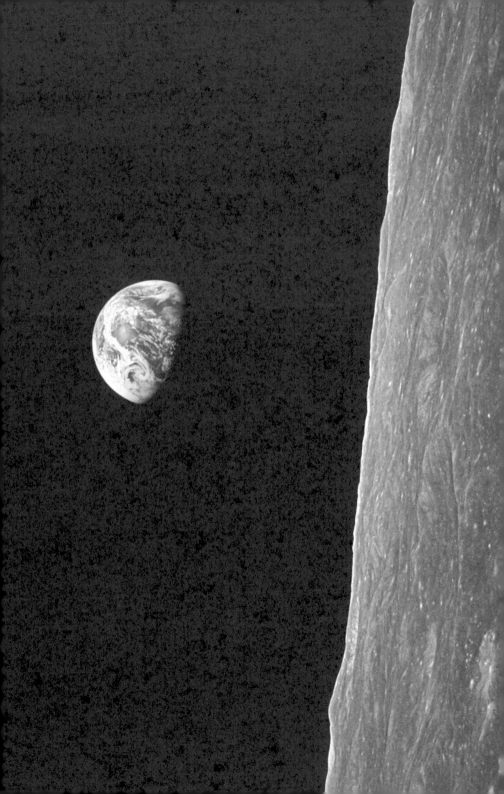

The World As I See It

How strange is the lot of us mortals! Each of us is here for a brief sojourn; for what purpose he knows not, though he sometimes thinks he senses it. But without deeper reflection one knows from daily life that one exists for other people—first of all for those upon whose smiles and well-being our own happiness is wholly dependent, and then for the many, unknown to us, to whose destinies we are bound by the ties of sympathy. A hundred times every day I remind myself that my inner and outer life are based on the labors of other men, living and dead, and that I must exert myself in order to give in the same measure as I have received and am still receiving. I am strongly drawn to a frugal life and am often oppressively aware that I am engrossing an undue amount of the labor of my fellowmen. I regard class distinctions as unjustified and, in the last resort, based on force. I also believe that a simple and unassuming life is good for everybody, physically and mentally.

I do not at all believe in human freedom in the philosophical sense. Everybody acts not only under external compulsion but also in accordance with

◀ **EARTHRISE FROM LUNAR ORBIT.** This spectacular image is perhaps the most famous in human history. Here we see it as viewed by the Apollo 8 crew instead of the usual way with the Earth above the lunar surface. Taken during the first manned orbits of the Moon before returning successfully to Earth, this photograph captures the fragile beauty of our planet. It helped inspire the environmental, human rights, and world peace movements by giving us a holistic vision of our blessed place in the universe.

inner necessity. Schopenhauer's saying, "A man can do as he wants, but not want what he wants," has been a very real inspiration to me since my youth; it has been a continual consolation in the face of life's hardships, my own and others', and an unfailing well-spring of tolerance. This realization mercifully mitigates the easily paralyzing sense of responsibility and it prevents us from taking ourselves and other people all too seriously; it is conducive to a view of life which, in particular, gives humor its due.

To inquire after the meaning or object of one's own existence or that of all creatures has always seemed to me absurd from an objective point of view. And yet everybody has certain ideals which determine the direction of his endeavors and his judgments. In this sense I have never looked upon ease and happiness as ends in themselves—this ethical basis I call the ideal of a pigsty. The ideals which have lighted my way, and time after time have given me new courage to face life cheerfully, have been Kindness, Beauty, and Truth. Without the sense of kinship with men of like mind, without the occupation with the objective world, the eternally unattainable in the field of art and scientific endeavors, life would have seemed to me empty. The trite objects of human efforts—possessions, outward success, luxury—have always seemed to me contemptible.

. . .

My passionate sense of social justice and social responsibility has always contrasted oddly with my pronounced lack of need for direct contact with other human beings and human communities. I am truly a "lone traveler" and have never belonged to my country, my home, my friends, or even my imme-diate family, with my whole heart; in the face of all these ties, I have never lost a sense of distance and a need for solitude—feelings which increase with the years. One becomes sharply aware, but without regret, of the limits of mutual understanding and consonance with other people. No doubt, such a person loses some of his innocence and unconcern; on the other hand, he is largely independent of the opinions, habits, and judgments of his fellows and avoids the temptation to build his inner equilibrium upon such insecure foundations.

. . .

My political ideal is democracy. Let every man be respected as an individual and no man idolized. It is an irony of fate that I myself have been the recipient of excessive admiration and reverence from my fellow-beings, through no fault, and no merit, of my own. The cause of this may well be the desire, unattainable for many, to understand the few ideas to which I have with my feeble powers attained through ceaseless struggle. I am quite aware that it is necessary for the achievement of the objective of an organization that one man should do the thinking and directing and generally bear the responsibility. But the led must not be coerced, they must be able to choose their leader.

An autocratic system of coercion, in my opinion, soon degenerates. For force always attracts men of low morality, and I believe it to be an invariable rule that tyrants of genius are succeeded by scoundrels. For this reason I have always been passionately opposed to systems such as we see in Italy and Russia today. The thing that has brought discredit upon the form of democracy as it exists in Europe today is not to be laid to the door of the democratic principle as such, but to the lack of stability of governments and to the impersonal character of the electoral system. I believe that in this respect the United States of America have found the right way. They have a President who is elected for a sufficiently long period and has sufficient powers really to exercise his responsibility. What I value, on the other hand, in the German political system is the more extensive provision that it makes for the individual in case of illness or need. The really valuable thing in the pageant of human life seems to me not the political state, but the creative, sentient individual, the personality; it alone creates the noble and sublime, while the herd as such remains dull in thought and dull in feeling.

This topic brings me to that worst outcrop of the herd life, the military system, which I abhor. That a man can take pleasure in marching in fours to the strains of a band is enough to make me despise him. He has only been given his big brain by mistake; unprotected spinal marrow was all he needed. This plague-spot of civilization ought to be abolished with all possible speed. Heroism by command, senseless violence, and all the loathsome nonsense that goes by the name of patriotism—how passionately I hate them! How

vile and despicable seems war to me! I would rather be hacked in pieces than take part in such an abominable business. My opinion of the human race is high enough that I believe this bogey would have disappeared long ago, had the sound sense of the peoples not been systematically corrupted by commercial and political interests acting through the schools and the Press.

The most beautiful experience we can have is the mysterious. It is the fundamental emotion which stands at the cradle of true art and true science. Whoever does not know it and can no longer wonder, no longer marvel, is as good as dead, and his eyes are dimmed. It was the experience of mystery—even if mixed with fear—that engendered religion. A knowledge of the existence of something we cannot penetrate, our perceptions of the profoundest reason and the most radiant beauty, which only in their most primitive forms are accessible to our minds—it is this knowledge and this emotion that constitute true religiosity; in this sense, and in this alone, I am a deeply religious man. I cannot conceive of a God who rewards and punishes his creatures, or

▲ **ABELL 2218, an enormous cluster of galaxies that resides in the constellation Draco some 2 billion light-years from Earth, is so massive that its gravitational field magnifies, brightens, and distorts the light of more distant objects. The phenomenon, known as a gravitational lens (predicted by Einstein's Theory of General Relativity) is evident by the arc-shaped patterns found throughout the image. These "arcs" are actually distorted images of very distant galaxies, which lie 5 to 10 times farther than Abell 2218.**

has a will of the kind that we experience in ourselves. Neither can I nor would I want to conceive of an individual that survives his physical death; let feeble souls, from fear or absurd egoism, cherish such thoughts. I am satisfied with the mystery of the eternity of life and with the awareness and a glimpse of the marvelous structure of the existing world, together with the devoted striving to comprehend a portion, be it ever so tiny, of the Reason that manifests itself in nature.

The human race is like a great uprooted tree, with its roots in the air. We must plant ourselves again in the universe.

—*D.H. Lawrence,* Lady Chatterley's Lover *(1928)*

Part II

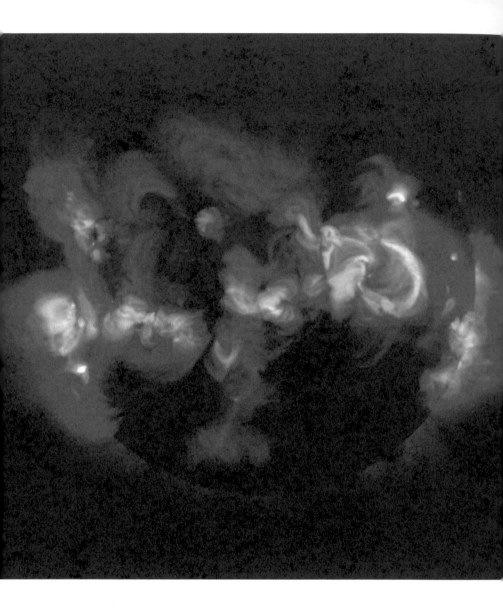

▲ **X-RAY SUN FROM YOHKOH SPACECRAFT.** In late 1991, the Yohkoh spacecraft obtained this spectacular image with its X-ray camera. The Sun was then near the top of its 11-year activity cycle. Along the Sun's equator the solar atmosphere was erupting with energetic regions. Views such as this trace the structure of magnetic fields because gas ionized by the Sun's million-degree corona flow only along the field lines.

Religion and Science

Everything that the human race has done and thought is concerned with the satisfaction of deeply felt needs and the assuagement of pain. One has to keep this constantly in mind if one wishes to understand spiritual movements and their development. Feeling and longing are the motive force behind all human endeavor and human creation, in however exalted a guise the latter may present themselves to us. Now what are the feelings and needs that have led men to religious thought and belief in the widest sense of the words? A little consideration will suffice to show us that the most varying emotions preside over the birth of religious thought and experience. With primitive man it is above all fear that evokes religious notions—fear of hunger, wild beasts, sickness, death. Since at this stage of existence understanding of causal connections is usually poorly developed, the human mind creates illusory beings more or less analogous to itself on whose wills and actions these fearful happenings depend. Thus one tries to secure the favor of these beings by carrying out actions and offering sacrifices which, according to the tradition handed down from generation to generation, propitiate them or make them well disposed toward a mortal. In this sense I am speaking of a religion of fear. This, though not created, is in an important degree stabilized by the formation of a special priestly caste which sets itself up as a mediator between the people and the beings they fear, and erects a hegemony on this basis. In many cases a leader or ruler or a privileged class whose position rests on other factors combines priestly functions with its secular authority in order to make the latter more secure; or the political rulers and the priestly caste make common cause in their own interests.

The social impulses are another source of the crystallization of religion. Fathers and mothers and the leaders of larger human communities are mortal and fallible. The desire for guidance, love, and support prompts men to form the social or moral conception of God. This is the God of Providence, who protects, disposes, rewards, and punishes; the God who, according to the limits of the believer's outlook, loves and cherishes the life of the tribe or of the human race, or even life itself; the comforter in sorrow and unsatisfied longing; he who preserves the souls of the dead. This is the social or moral conception of God.

The Jewish scriptures admirably illustrate the development from the religion of fear to moral religion, a development continued in the New Testament. The religions of all civilized peoples, especially the peoples of the Orient, are primarily moral religions. The development from a religion of fear to moral religion is a great step in peoples' lives. And yet, that primitive religions are based entirely on fear and the religions of civilized peoples purely on morality is a prejudice against which we must be on our guard. The truth is that all religions are a varying blend of both types, with this differentiation: that on the higher levels of social life the religion of morality predominates.

Common to all these types is the anthropomorphic character of their conception of God. In general, only individuals of exceptional endowments, and exceptionally high-minded communities, rise to any considerable extent above this level. But there is a third stage of religious experience which belongs to all of them, even though it is rarely found in a pure form: I shall call it cosmic religious feeling. It is very difficult to elucidate this feeling to anyone who is entirely without it, especially as there is no anthropomorphic conception of God corresponding to it.

The individual feels the futility of human desires and aims and the sublimity and marvelous order which reveal themselves both in nature and in the world of thought. Individual existence impresses him as a sort of prison and he wants to experience the universe as a single significant whole. The beginnings of cosmic religious feeling already appear at an early stage of development, e.g., in many of the Psalms of David and in some of the Prophets.

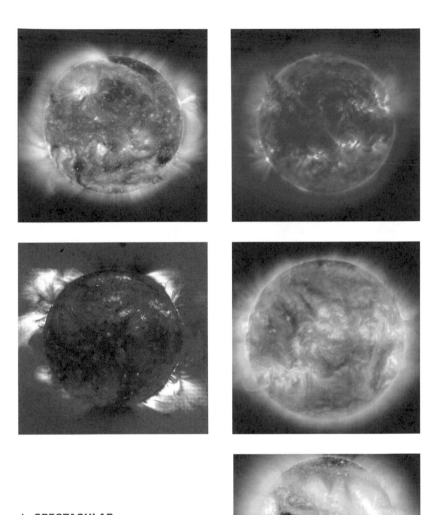

▲ **SPECTACULAR VIEWS OF OUR SUN** depicted in different wavelengths of light. These images were made by the EIT camera onboard the SOHO spacecraft, a space observatory which can continuously observe the Sun.

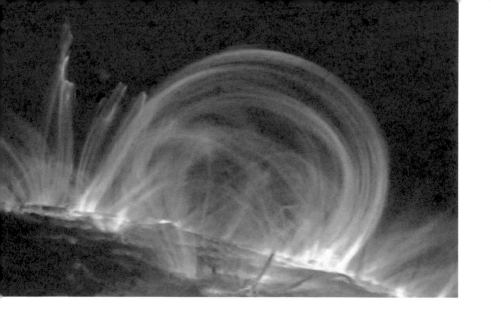

▲ **CORONAL LOOPS.** Giant fountains of fast-moving, multimillion-degree gas in the outermost atmosphere of the Sun.

Buddhism, as we have learned especially from the wonderful writings of Schopenhauer, contains a much stronger element of this.

The religious geniuses of all ages have been distinguished by this kind of religious feeling, which knows no dogma and no God conceived in man's image; so that there can be no church whose central teachings are based on it. Hence it is precisely among the heretics of every age that we find men who were filled with this highest kind of religious feeling and were in many cases regarded by their contemporaries as atheists, sometimes also as saints. Looked at in this light, men like Democritus, Francis of Assisi, and Spinoza are closely akin to one another.

How can cosmic religious feeling be communicated from one person to another, if it can give rise to no definite notion of a God and no theology? In my view, it is the most important function of art and science to awaken this feeling and keep it alive in those who are receptive to it.

We thus arrive at a conception of the relation of science to religion very different from the usual one. When one views the matter historically, one is inclined to look upon science and religion as irreconcilable antagonists, and for a very obvious reason. The man who is thoroughly convinced of the universal operation of the law of causation cannot for a moment entertain the idea of a being who interferes in the course of events—provided, of course, that he takes the hypothesis of causality really seriously. He has no use for the religion of fear and equally little for social or moral religion. A God who rewards and punishes is inconceivable to him for the simple reason that a man's actions are determined by necessity, external and internal, so that in God's eyes he cannot be responsible, any more than an inanimate object is responsible for the motions it undergoes. Science has therefore been charged with undermining morality, but the charge is unjust. A man's ethical behavior should be based effectually on sympathy, education, and social ties and needs; no religious basis is necessary. Man would indeed be in a poor way if he had to be restrained by fear of punishment and hope of reward after death.

It is therefore easy to see why the churches have always fought science and persecuted its devotees. On the other hand, I maintain that the cosmic religious feeling is the strongest and noblest motive for scientific research. Only those who realize the immense efforts and, above all, the devotion without which pioneer work in theoretical science cannot be achieved are able to grasp the strength of the emotion out of which alone such work, remote as it is from the immediate realities of life, can issue. What a deep conviction of the rationality of the universe and what a yearning to understand, were it but a feeble reflection of the mind revealed in this world, Kepler and Newton must have had to enable them to spend years of solitary labor in disentangling the principles of celestial mechanics! Those whose acquaintance with scientific research is derived chiefly from its practical results easily develop a completely false notion of the mentality of the men who, surrounded by a skeptical world, have shown the way to kindred spirits scattered wide through the world and the centuries. Only one who has devoted his life to similar ends can have a vivid realization of what has inspired these men and given them the strength to remain true to their purpose in spite of countless failures. It is cosmic religious feeling that gives a man such strength. A contemporary has said, not unjustly, that in this materialistic age of ours the serious scientific workers are the only profoundly religious people.

. . .

One thing I have learned in a long life: that all our science, measured against reality, is primitive and childlike—and yet it is the most precious thing we have.

. . .

The cosmic religious experience is the strongest and noblest mainspring of scientific research.

. . .

Science is no more than the purification of daily thoughts.

. . .

By furthering logical thought and a logical attitude, science can diminish the amount of superstition in the world. There is no doubt that all but the crudest scientific work is based on a firm belief—akin to religious feeling—in the rationality and comprehensibility of the world.

. . .

▲ **Illustration of a CME Particle Cloud blasted from the Sun impacting Earth and creating an aurora.**

Science without religion is lame, religion without science is blind.

. . .

The further the spiritual evolution of mankind advances, the more certain it seems to me that the path to genuine religiosity does not lie through the fear of life, and the fear of death, and blind faith, but through striving after rational knowledge.

Morality and Values

I do not believe that the basic ideas of the theory of relativity can lay claim to a relationship with the religious spheres that is different from that of scientific knowledge in general. I see this connection in the fact that profound interrelationships in the objective world can be comprehended through simple logical concepts. To be sure, in the theory of relativity this is the case in particularly full measure.

The religious feeling engendered by experiencing the logical comprehensibility of profound interrelations is of a somewhat different sort from the feeling that one usually calls religious. It is more a feeling of awe at the scheme that is manifested in the material universe. It does not lead us to take the step of fashioning a god-like being in our own image—a personage who makes demands of us and who takes an interest in us as individuals. There is in this neither a will nor a goal, nor a must, but only sheer being. For this reason, people of our type see in morality a purely human matter, albeit the most important in the human sphere.

·   ·   ·

Our time is distinguished by wonderful achievements in the field of scientific understanding and the technical application of those insights. Who would not be cheered by this? But let us not forget that knowledge and skills alone

◄ **THE NEBULA AROUND ANTARES.** Antares is a giant among stars, and like giants in general, very rare. It has spent most of its short life as a highly luminous blue supergiant star, its mass obliging it to consume its vast store of nuclear fuel (hydrogen) very rapidly.

cannot lead humanity to a happy and dignified life. Humanity has every reason to place the proclaimers of high moral standards and values above the discoverers of objective truth. What humanity owes to personalities like Buddha, Moses, and Jesus ranks for me higher than all the achievements of the enquiring and constructive mind.

What these blessed men have given us we must guard and try to keep alive with all our strength if humanity is not to lose its dignity, the security of its existence, and its joy in living.

. . .

It is the moral qualities of its leading personalities that are perhaps of even greater significance for a generation and for the course of history than purely intellectual accomplishments. Even these latter are, to a far greater degree than is commonly credited, dependent on the stature of character.

. . .

It is right in principle that those should be best loved who have contributed most to the elevation of the human race and human life. But if one goes on to ask who they are, one finds oneself in no inconsiderable difficulties. In the case of political, and even of religious, leaders it is often very doubtful whether they have done more good or harm. Hence I most seriously believe that one does people the best service by giving them some elevating work to do and indirectly elevating them. This applies most of all to the great artist, but also in a lesser degree to the scientist. To be sure, it is not the fruits of scientific research that elevate a man and enrich his nature, but the urge to understand, the intellectual work, creative or receptive. Thus, it would surely be inappropriate to judge the value of the Talmud by its intellectual fruits.

. . .

The most important human endeavor is the striving for morality in our actions. Our inner balance and even our very existence depend on it. Only morality in our actions can give beauty and dignity to life.

To make this a living force and bring it to clear consciousness is perhaps the foremost task of education.

The foundation of morality should not be made dependent on myth nor

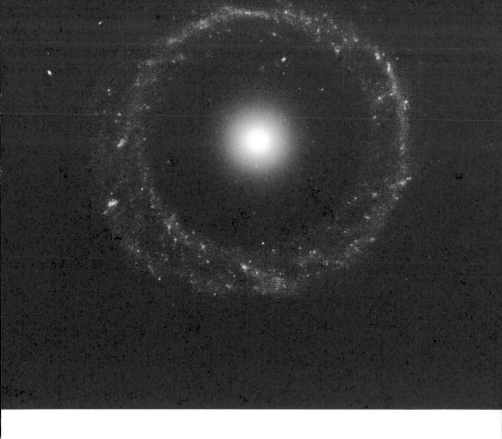

▲ **HOAG'S OBJECT.**
A nearly perfect ring of hot,
blue stars pinwheels about
the yellow nucleus of an
unusual galaxy known as
Hoag's Object. The entire
galaxy is about 120,000
light-years wide, which
is slightly larger than our
Milky Way galaxy.

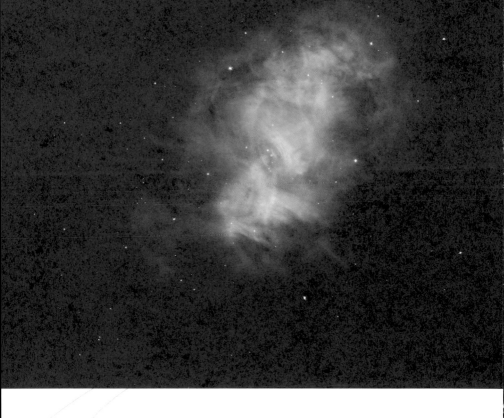

▲ **CASSIOPEIA A IN X-RAY.** In August of 1999, NASA released an image of Cassiopeia A, a supernova remnant revealed in never-before-seen X-ray detail. The "Cas A" image shows remarkable structure in the debris of a gigantic stellar explosion, as well as an enigmatic source in the center, which could be a rapidly spinning neutron star or black hole.

tied to any authority lest doubt about the myth or about the legitimacy of the authority imperil the foundation of sound judgement and action

·　·　·

Looked at from a simple human point of view, moral conduct does not mean merely a stern demand to renounce some of the desired joys of life, but rather a sociable interest in a happier lot for all men.

This conception implies one requirement above all—that every individual should have the opportunity to develop the gifts which may be latent in him. Alone in that way can the individual obtain the satisfaction to which he is justly entitled; and alone in that way can the community achieve its richest flowering. For everything that is really great and inspiring is created by the individual who can labour in freedom. Restriction is justified only in so far as it may be needed for the security of existence.

There is one other thing which follows from that conception—that we must not only tolerate differences between individuals and between groups, but we should indeed welcome them and look upon them as an enriching of our existence. That is the essence of all true tolerance; without tolerance in this widest sense there can be no question of true morality.

·　·　·

The true value of a human being is determined primarily by the measure and the sense in which he has attained to liberation from the self.

·　·　·

I cannot conceive of a personal God who would directly influence the actions of individuals, or would directly sit in judgement on creatures of his own creation. I cannot do this in spite of the fact that mechanistic causality has, to a certain extent, been placed in doubt by modern science.

My religiosity consists in a humble admiration of the infinitely superior spirit that reveals itself in the little that we, with our weak and transitory understanding, can comprehend of reality. Morality is of the highest importance—but for us, not for God.

·　·　·

Never do anything against your conscience even if the state demands it.

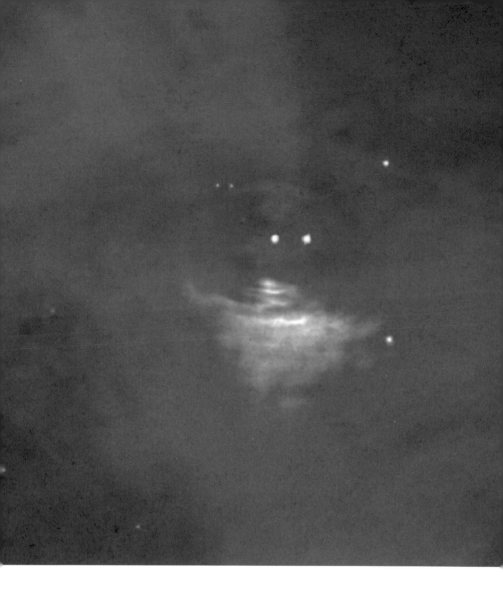

▲ **CRAB NEBULA.** This eerie photograph is a close-up of the inner parts of the Crab nebula. The Crab pulsar (seen here at the left of the pair of stars near the center of the frame) is the collapsed core of the exploded star. The pulsar itself is a rapidly rotating neutron star, an object only about 10 kilometers (6 miles) across, but containing more mass than our Sun.

Moral Decay

All religions, arts and sciences are branches of the same tree. All these aspirations are directed toward ennobling man's life, lifting it from the sphere of mere physical existence and leading the individual toward freedom. It is no mere chance that our older universities have developed from clerical schools. Both churches and universities—insofar as they live up to their function—serve the ennoblement of the individual. They seek to fulfill this great task by spreading moral and cultural understanding, renouncing the use of brute force.

The essential unity of ecclesiastical and secular institutions was lost during the 19th century, to the point of senseless hostility. Yet there never was any doubt as to the striving of culture. No one doubted the sacredness of the goal. It was the approach that was disputed.

The political and economic conflicts and complexities of the last few decades have brought before our eyes dangers which even the darkest pessimists of the last century did not dream of. The injunctions of the Bible concerning human conduct were then accepted by believer and infidel alike as self-evident demands for the individuals and society. No one would have taken seriously who failed to acknowledge the quest for objective truth and knowledge as man's highest and eternal aim.

Yet today we must recognize with horror that these pillars of civilized human existence have lost their firmness. Nations that once ranked high bow down before tyrants who dare openly to assert: Right is that which serves us! The quest for truth for its own sake has no justification and is not to be tolerated. Arbitrary rule, oppression, persecution of individuals, faiths and communities are openly practiced in those countries and accepted as justifiable or inevitable.

And the rest of the world has slowly grown accustomed to these symptoms of moral decay. One misses the elementary reaction against injustice and for justice—that reaction which in the long run represents man's only protection against a relapse into barbarism. I am firmly convinced that the passionate will for justice and truth has done more to improve man's condition than calculating political shrewdness which in the long run only breeds general distrust. Who can doubt that Moses was a better leader of humanity than Machiavelli?

During the War someone tried to convince a great Dutch scientist that might went before right in the history of man. "I cannot disprove the accuracy of your assertion," he replied, "but I do know that I should not care to live in such a world!"

Let us think, feel and act like this man, refusing to accept fateful compromise. Let us not even shun the fight when it is unavoidable to preserve right and the dignity of man. If we do this we shall soon return to conditions that will allow us to rejoice in humanity.

.   .   .

Anger dwells only in the bosom of fools.

.   .   .

Great spirits have always encountered violent opposition from mediocre minds.

.   .   .

It is my contention that killing under the cloak of war is nothing but an act of murder.

▶ **WITCHHEAD NEBULA. Appearing to cackle with wicked delight, the Witchhead nebula is actually the remnants of an ancient supernova explosion located about 1,000 light-years away in the constellation Eridanus. Its gases glow from the light of the supergiant star Rigel in the neighboring Orion nebula.**

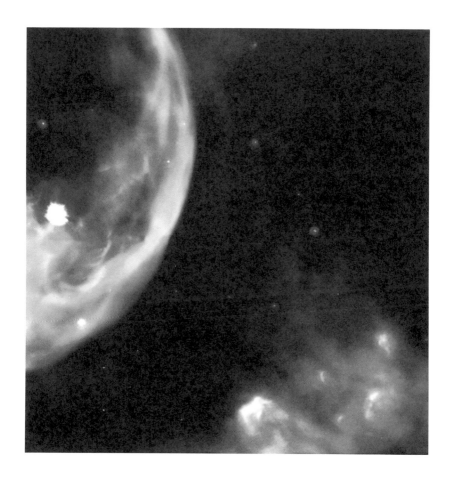

▲ **BUBBLE NEBULA, NGC 7635.** An expanding shell of glowing gas surrounds a hot, massive star in our Milky Way galaxy. This shell is being shaped by strong stellar winds of material and radiation produced by the bright star at the left, which is 10 to 20 times more massive than our Sun. The nebula is 10 light-years across, more than twice the distance from Earth to the nearest star. Only part of the bubble is visible in this image.

Christianity and Judaism

The highest principles for our aspirations and judgments are given to us in the Jewish-Christian religious tradition. It is a very high goal which, with our weak powers, we can reach only very inadequately, but which gives a sure foundation to our aspirations and valuations. If one were to take that goal out of its religious form and look merely at its purely human side, one might state it perhaps thus: free and responsible development of the individual, so that he may place his powers freely and gladly in the service of all mankind.

There is no room in this for the divinization of a nation, of a class, let alone of an individual. Are we not all children of one father, as it is said in religious language? Indeed, even the divinization of humanity, as an abstract totality, would not be in the spirit of that ideal. It is only to the individual that a soul is given. And the high destiny of the individual is to serve rather than rule, or to impose himself in any other way.

.　　.　　.

If one purges the Judaism of the Prophets and Christianity as Jesus Christ taught it, of all subsequent additions, especially those of the priests, one is left with a teaching which is capable of curing all the social ills of humanity.

It is the duty of every man of good will to strive steadfastly in his own little world to make this teaching of pure humanity a living force, so far as he can. If he makes an honest attempt in this direction without being crushed and trampled under foot by his contemporaries, he may consider himself and the community to which he belongs lucky.

.　　.　　.

If the believers of the present-day religions would earnestly try to think and act in the spirit of the founders of these religions then no hostility on the basis of religion would exist among the followers of the different faiths. Even the conflicts in the realm of religion would be exposed as insignificant.

. . .

For me the essence of religion is to be able to get under the skin of another human being, to rejoice in his joy and suffer his pain.

▶ **WOLF-RAYET BINARY AND NEBULA.**
Two young stars power these colorful interstellar gas clouds. These objects are located in the Large Magellanic Cloud (LMC), the largest satellite galaxy to our own Milky Way galaxy. One of the stars in the central binary is an enigmatic Wolf-Rayet star while the other is a massive O star. Wolf-Rayet stars have some of the hottest surfaces in the universe.

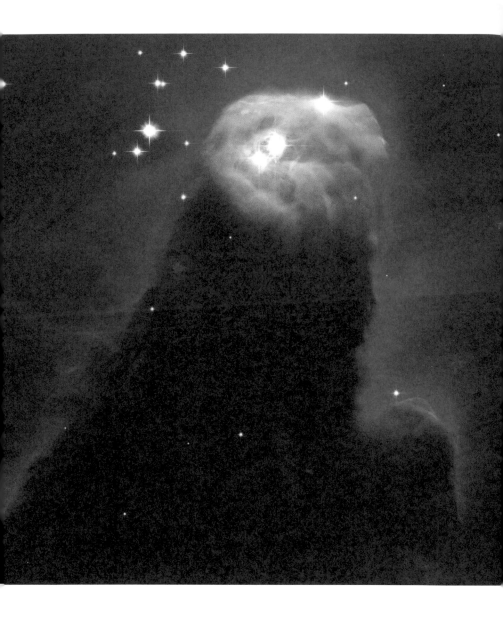

▲ **CONE NEBULA.** Cones, pillars, and majestic flowing shapes abound in stellar nurseries where natal clouds of gas and dust are buffeted by energetic winds from newborn stars. The Cone nebula within the bright galactic star-forming region NGC 2264, was captured in this close-up view. The Cone nebula's reddish veil is produced by glowing hydrogen gas.

God

About God, I cannot accept any concept based on the authority of the Church. As long as I can remember, I have resented mass indoctrination. I do not believe in fear of life, in the fear of death, in blind faith. I cannot prove to you that there is no personal God, but if I were to speak of him, I would be a liar. I do not believe in the God of theology who rewards and punishes evil. My God created laws that take care of that. His universe is not ruled by wishful thinking, but by immutable laws.

.　　.　　.

Whatever there is of God and goodness in the universe, it must work itself out and express itself through us. We cannot stand aside and let God do it.

.　　.　　.

Before God we are all equally wise—equally foolish.

.　　.　　.

I see a pattern. But my imagination cannot picture the maker of that pattern. I see the clock. But I cannot envisage the clock-maker. The human mind is unable to conceive of the four dimensions. How can it conceive of a God, before whom a thousand years and a thousand dimensions are as one?

.　　.　　.

I believe in Spinoza's God, who reveals himself in a harmony among all people, not in a God who worries about the destiny and actions of man.

.　　.　　.

What I'm really interested in is whether God could have made the world in a different way; that is, whether the necessity of logical simplicity leaves any freedom at all.

.　　.　　.

God is subtle, but he is not malicious.

.   .   .   .

I want to know how God created this world. I am not interested in this or
that phenomenon, in the spectrum of this or that element. I want to know His
thoughts, the rest are details.

▲ **THE SOMBRERO GALAXY, MESSIER 104** (M 104), is one of the
universe's most stately and photogenic galaxies. Encircled by the thick
dust lanes comprising the spiral structure of the galaxy, its hallmark
is a brilliant white, bulbous core. Named the Sombrero because of its
resemblance to the broad rim and high-topped Mexican hat, it lies at
the southern edge of the rich Virgo cluster of galaxies and is one of the
most massive objects in that group, equivalent to 800 billion suns.

Prayer

Scientific research is based on the idea that everything that takes place is determined by laws of nature, and therefore this holds for the actions of people. For this reason, a research scientist will hardly be inclined to believe that events could be influenced by a prayer, i.e. by a wish addressed to a supernatural Being.

However, it must be admitted that our actual knowledge of these laws is only imperfect and fragmentary, so that, actually, the belief in the existence of basic all-embracing laws in Nature also rests on a sort of faith. All the same this faith has been largely justified so far by the success of scientific research.

But, on the other hand, every one who is seriously involved in the pursuit of science becomes convinced that a spirit is manifest in the laws of the Universe—a spirit vastly superior to that of man, and one in the face of which we with our modest powers must feel humble. In this way the pursuit of science leads to a religious feeling of a special sort, which is indeed quite different from the religiosity of someone more naive.

◀ **NGC 602 N 90. Dwarf galaxies such as the Small Magellanic Cloud, with significantly fewer stars compared to our own galaxy, are considered to be the primitive building blocks of larger galaxies. At the heart of the star-forming region, lies star cluster NGC 602. The high-energy radiation blazing out from the hot young stars is sculpting the inner edge of the outer portions of the nebula, slowly eroding it away and eating into the material beyond.**

▲ **NGC 604** is a vast nebula which lies in the neighboring spiral galaxy M 33, located 2.7 million light-years away in the constellation Triangulum. New stars are being born here in a spiral arm of the galaxy. At the heart of NGC 604 are over 200 hot stars, much more massive than our Sun (15 to 60 solar masses). Their light also highlights the nebula's three-dimensional shape, like a lantern in a cavern.

Mysticism

The mystical trend of our time, which shows itself particularly in the rampant growth of the so-called Theosophy and Spiritualism, is for me no more than a symptom of weakness and confusion.

Since our inner experiences consist of reproductions and combinations of sensory impressions, the concept of a soul without a body seems to me to be empty and devoid of meaning.

.　　.　　.

Body and soul are not two different things, but only two different ways of perceiving the same thing. Similarly, physics and psychology are only different attempts to link our experiences together by way of systematic thought.

.　　.　　.

I do not believe in immortality of the individual, and I consider ethics to be an exclusively human concern with no superhuman authority behind it.

.　　.　　.

I have never imputed to Nature a purpose or a goal, or anything that could be understood as anthropomorphic.

What I see in Nature is a magnificent structure that we can comprehend only very imperfectly, and that must fill a thinking person with a feeling of "humility." This is a genuinely religious feeling that has nothing to do with mysticism.

The first peace, which is the most important,
is that which comes within the souls of the people
when they realize their relationship, their oneness
with the universe and all its powers, and when they
realize that at the center of the universe dwells
the Great Spirit, and that this center is really
everywhere, it is within each of us.

—*Black Elk (1863-1950)*

Part III

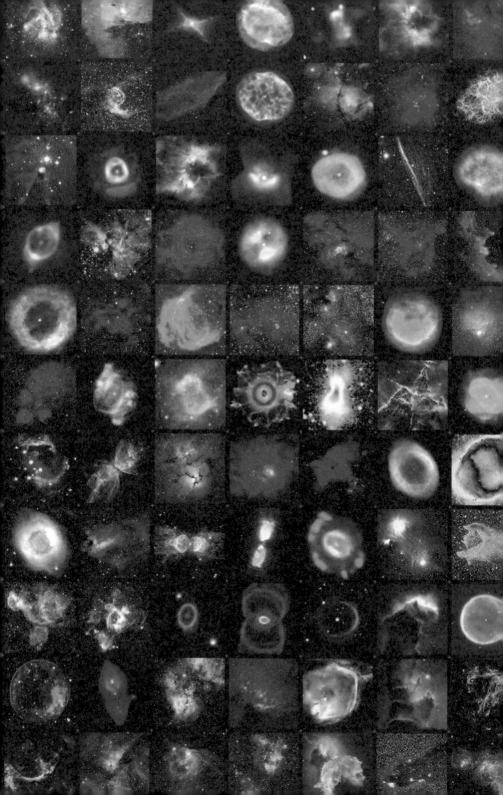

The Individual

I believe in the brotherhood of man and the uniqueness of the individual. But if you ask me to prove what I believe, I can't. You know them to be true but you could spend a whole lifetime without being able to prove them. The mind can proceed only so far upon what it knows and can prove. There comes a point where the mind takes a higher plane of knowledge, but can never prove how it got there. All great discoveries have involved such a leap.

. . .

It is only to the individual that a soul is given.

. . .

Everything that is really great and inspiring is created by the individual who can labor in freedom.

. . .

Valuable achievement can sprout from human society only when it is sufficiently loose to make possible the free development of an individual's abilities.

◀ **Montage of Planetary Nebulae.**

Morals and Emotions

We all know, from what we experience with and within ourselves, that our conscious acts spring from our desires and our fears. Intuition tells us that that is true also of our fellows and of the higher animals. We all try to escape pain and death, while we seek what is pleasant. We all are ruled in what we do by impulses; and these impulses are so organized that our actions in general serve for our self-preservation and that of the race. Hunger, love, pain, fear are some of those inner forces which rule the individuals instinct for self-preservation. At the same time, as social beings, we are moved in the relations with our fellow beings by such feelings as sympathy, pride, hate, need for power, pity, and so on. All these primary impulses, not easily described in words, are the springs of man's actions. All such action would cease if those powerful elemental forces were to cease stirring within us.

Though our conduct seems so very different from that of the higher animals, the primary instincts are much alike in them and in us. The most evident difference springs from the important part which is played in man by a relatively strong power of imagination and by the capacity to think, aided as it is by language and other symbolical devices. Thought is the organizing factor

◄ **THE WHIRLPOOL GALAXY, M 51, has been one of the most photogenic galaxies in amateur and professional astronomy. This celestial beauty is studied extensively in a range of wavelengths by large ground- and space-based observatories. The companion's gravitational pull is triggering star formation in the main galaxy, as seen in brilliant detail by numerous, luminous clusters of young and energetic stars.**

in man, intersected between the causal primary instincts and the resulting actions. In that way imagination and intelligence enter into our existence in the part of servants of the primary instincts. But their intervention makes our acts to serve ever less merely the immediate claims of our instincts. Through them the primary instinct attaches itself to ends which become ever more distant. The instincts bring thought into action, and thought provokes intermediary actions inspired by emotions which are likewise related to the ultimate end. Through repeated performance, this process brings it about that ideas and beliefs acquire and retain a strong effective power even after the ends which gave them that power are long forgotten. In abnormal cases of such intensive borrowed emotions, which cling to objects emptied of their erstwhile effective meaning, we speak of fetishism.

Yet the process which I have indicated plays a very important part also in ordinary life. Indeed there is no doubt that to this process—which one may describe as a spiritualizing of the emotions and of thought—that to it man owes the most subtle and refined pleasures of which he is capable: the pleasure in the beauty of artistic creation and of logical trains of thought.

As far as I can see, there is one consideration which stands at the threshold of all moral teaching. If men as individuals surrender to the call of their elementary instincts, avoiding pain and seeking satisfaction only for their own selves, the result for them all taken together must be a state of insecurity, of fear, and of promiscuous misery. If, besides that, they use their intelligence from an individualist, i.e., a selfish standpoint, building up their life on the

▶ **THACKERAY'S GLOBULES IN IC 2944.** Strangely glowing dark clouds float serenely in this remarkable and beautiful image taken with the Hubble Space Telescope. These dense, opaque dust clouds (known as "globules") are silhouetted against nearby bright stars in the busy star-forming region, IC 2944. Astronomers still know very little about their origin and nature, except that they are generally associated with areas of star formation, called "HII regions" due to the presence of hydrogen gas. These stars are much hotter and much more massive than our Sun.

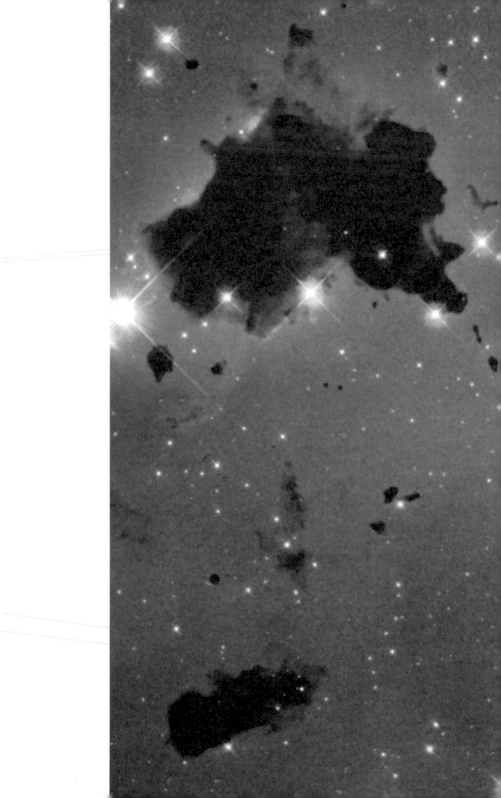

illusion of a happy unattached existence, things will be hardly better. In comparison with the other elementary instincts and impulses, the emotions of love, of pity and of friendship are too weak and too cramped to lead to a tolerable state of human society.

The solution of this problem, when freely considered, is simple enough, and it seems also to echo from the teachings of the wise men of the past always in the same strain: All men should let their conduct be guided by the same principles; and those principles should be such, that by following them there should accrue to all as great a measure as possible of security and satisfaction, and as small a measure as possible of suffering.

Of course, this general requirement is much too vague that we should be able to draw from it with confidence specific rules to guide the individuals in their actions. And indeed, these specific rules will have to change in keeping with changing circumstances. If this were the main difficulty that stands in the way of that keen conception, the millenary fate of man would have been incomparably happier than it actually was, or still is. Man would not have killed man, tortured each other, exploited each other by force and by guile.

The real difficulty, the difficulty which has baffled the sages of all times, is rather this: how can we make our teaching so potent in the emotional life of man, that its influence should withstand the pressure of the elemental psychic forces in the individual? We do not know, of course, if the sages of the past have really asked themselves this question, consciously and in this form; but we do know how they have tried to solve the problem.

Long before men were ripe, namely, to be faced with such a universal moral attitude, fear of the dangers of life had led them to attribute to various imaginary personal beings, not physically tangible, power to release those natural forces which men feared or perhaps welcomed. And they believed that those beings, which everywhere dominated their imagination, were psychically made in their own image, but were endowed with superhuman powers. These were the primitive precursors of the idea of God. Sprung in the first place from the fears which filled man's daily life, the belief in the existence of such beings, and in their extraordinary powers, has had so strong an influence on men and their conduct, that it is difficult for us to imagine. Hence it is not surprising that those who set out to establish the moral idea, as embracing all men equally, did so by linking it closely with religion. And the fact that those moral claims were the same for all men, may have had much to do with the development of mankind's religious culture from polytheism to monotheism.

The universal moral idea thus owed its original psychological potency to that link with religion. Yet in another sense that close association was fatal for the moral idea. Monotheistic religion acquired different forms with various peoples and groups. Although those differences were by no means fundamental, yet they soon were felt more strongly than the essentials that were common. And in that way religion often caused enmity and conflict, instead of binding mankind together with the universal moral idea.

Then came the growth of the natural sciences, with their great influence on thought and practical life, weakening still more in modern times the religious sentiment of the peoples. The causal and objective mode of thinking—though not necessarily in contradiction with the religious sphere—leaves in most people little room for a deepening religious sense. And because of the traditional close link between religion and morals, that has brought with it, in the last hundred years or so, a serious weakening of moral thought and sentiment. That, to my mind, is a main cause for the barbarization of political ways in our time. Taken together with the terrifying efficiency of the new technical means, the barbarization already forms a fearful threat for the civilized world.

Needless to say, one is glad that religion strives to work for the realization of the moral principle. Yet the moral imperative is not a matter for church and religion alone, but the most precious traditional possession of all mankind. Consider from this standpoint the position of the Press, or of the schools with their competitive method! Everything is dominated by the cult of efficiency and of success and not by the value of things and merit in relation to the moral ends of human society. To that must be added the moral deterioration resulting from a ruthless economic struggle. The deliberate nurturing of the moral sense also outside the religious sphere, however, should help also in this, to lead men to look upon social problems as so many opportunities for joyous service towards a better life. For looked at from a simple human point of view, moral conduct does not mean merely a stern demand to renounce some of the desired joys of life, but rather a sociable interest in a happier lot for all men.

This conception implies one requirement above all—that every individual should have the opportunity to develop the gifts which may be latent in him. Alone in that way can the individual obtain the satisfaction to which he is justly entitled; and alone in that way can the community achieve its richest flowering. For everything that is really great and inspiring is created by the individual who can labour in freedom. Restriction is justified only in so far as it may be needed for the security of existence.

There is one other thing which follows from that conception—that we must not only tolerate differences between individuals and between groups, but we should indeed welcome them and look upon them as an enriching of our existence. That is the essence of all true tolerance; without tolerance in this widest sense there can be no question of true morality.

Morality in the sense here briefly indicated is not a fixed and stark system. It is rather a standpoint from which all questions which arise in life could and should be judged. It is a task never finished, something always present to guide our judgment and to inspire our conduct. Can you imagine that any man filled with this ideal could be content: ?

Were he to receive from his fellow men a much greater return in goods and services than most other men ever receive?

Were his country, because it feels itself for the time being militarily secure, to stand aloof from the aspiration to create a supra-national system of security and justice?

Could he look on passively, or perhaps even with indifference, when elsewhere in the world innocent people are being brutally persecuted, deprived of their rights or even massacred?

To ask these questions is to answer them!

▶ **THE ORION NEBULA from CFHT.**

Of Wealth

I am absolutely convinced that no wealth in the world can help humanity forward, even in the hands of the most devoted worker in this cause. The examples of great and pure individuals is the only thing that can lead us to noble thoughts and deeds. Money only appeals to selfishness and irresistibly invites abuse.

Can anyone imagine Moses, Jesus or Gandhi armed with the money-bags of Carnegie?

. . .

Not everything that counts can be counted and not everything that can be counted counts.

◄ **CHAMELEON I COMPLEX, NGC 3195.** In the constellation Chameleon are a group of photogenic nebulae, predominately visible in the skies south of the Earth's equator. Dark molecular clouds and bright planetary nebula NGC 3195 can be found toward Chameleon. On the lower right, a dark molecular cloud blocks the stars behind it. The light from these objects takes hundreds of years to reach us.

I simply can't build my hopes on a foundation
of confusion, misery and death . . . I think . . .
peace and tranquility will return again.

—*Anne Frank*

Part IV

▲ **A NUCLEAR EXPLOSION** in the Mururoa Atoll, in French Polynesia,
on Sunday 6 March 1971 03:59 PM. From 1945 until 1998, there have
been over 2,000 nuclear tests worldwide. Nuclear weapons in today's
arsenals have a total yield of over 200,000 Hiroshima bombs.

The Menace of
Mass Destruction

Everyone is aware of the difficult and menacing situation in which human society—shrunk into one community with a common fate—finds itself, but only a few act accordingly. Most people go on living their everyday life: half-frightened, half indifferent, they behold the ghostly tragic-comedy that is being performed on the international stage before the eyes and ears of the world. But on that stage, on which the actors under the floodlights play their ordained parts, our fate of tomorrow, life and death of the nations, is being decided.

It would be different if the problem were not one of things made by Man himself, such as the atomic bomb and other means of mass destruction equally menacing all peoples. It would be different, for instance, if an epidemic of bubonic plague were threatening the entire world. In such a case conscientious and expert persons would be brought together and they would work out an intelligent plan to combat the plague. After having reached agreement upon the right ways and means, they would submit their plan to the governments. Those would hardly raise serious objections but rather agree speedily on the measures to be taken. They certainly would never think of trying to handle the matter in such a way that their own nation would be spared whereas the next one would be decimated.

But could not our situation be compared to one of a menacing epidemic? People are unable to view this situation in its true light, for their eyes are blinded by passion. General fear and anxiety create hatred and aggressiveness. The adaptation to warlike aims and activities has corrupted the mentality of man; as a result, intelligent, objective and humane thinking has hardly any effect and is even suspected and persecuted as unpatriotic.

There are, no doubt, in the opposite camps enough people of sound judgment and sense of justice who would be capable and eager to work out together a solution for the factual difficulties. But the efforts of such people are hampered by the fact that it is made impossible for them to come together for informal discussions. I am thinking of persons who are accustomed to the objective approach to a problem and who will not be confused by exaggerated nationalism or other passions. This forced separation of the people of both camps I consider one of the major obstacles to the achievement of an acceptable solution of the burning problem of international security.

As long as contact between the two camps is limited to the official negotiations I can see little prospect for an intelligent agreement being reached, especially since considerations of national prestige as well as the attempt to talk out of the window for the benefit of the masses are bound to make reasonable progress almost impossible. What one party suggests officially is for that reason alone suspected and even made unacceptable to the other. Also behind all official negotiations stands—though veiled—the threat of naked power. The official method can lead to success only after spade-work of an informal nature has prepared the ground; the conviction that a mutually satisfactory solution can be reached must be gained first; then the actual negotiations can get under way with a fair promise of success.

We scientists believe that what we and our fellow-men do or fail to do within the next few years will determine the fate of our civilization. And we consider it our task untiringly to explain this truth, to help people realize all that is at stake, and to work, not for appeasement, but for understanding and ultimate agreement between peoples and nations of different views.

·　　·　　·

Peace cannot be kept by force. It can only be achieved by understanding.

·　　·　　·

I know not how the Third World War will be fought, but I can tell you what they will use in the Fourth—rocks!

·　　·　　·

This is not a comedy. It is the greatest tragedy of modern times, despite the cap and bells and buffoonery. We should be standing on rooftops... denouncing this as a travesty! ...

.　　.　　.

If you want peace... ask the workers to refuse to manufacture and transport military weapons, and to refuse to serve in the military. Governments could go on talking now to doomsday.

.　　.　　.

The unleashed power of the atom bomb has changed everything except our modes of thinking, and thus we drift toward unparalleled catastrophe.

.　　.　　.

If only I had known, I should have become a watchmaker.

.　　.　　.

The solution to this problem lies in the heart of mankind.

.　　.　　.

We must inoculate our children against militarism, by educating them in the spirit of pacifism... Our schoolbooks glorify war and conceal its horrors. They indoctrinate children with hatred. I would teach peace rather than war, love rather than hate.

.　　.　　.

Concern for man himself must always be the chief interest of all technical endeavors... to assure that the results of our scientific thinking may be a blessing to mankind and not a curse. Never forget this in the midst of your diagrams and equations.

.　　.　　.

Theoretically there is no authority whose decisions and statements can claim to be the truth. Is that time forever passed, when aroused by his inner freedom and the independence of his thinking and his work, the scientist has the opportunity of enlightening and enriching the life of his fellow human beings? Has he not forgotten about his responsibility and dignity as a scientist?

▲ **MILKY WAY, PLANET JUPITER,** Star Arcturus,
and the Big Dipper over the Grand Tetons.

World Peace

If we have courage to decide ourselves for peace, we will have peace... We are not engaged in a play but in a condition of utmost danger to existence. If you are not firmly decided to resolve things in a peaceful way, you will never come to a peaceful solution.

.   .   .

I hold that mankind is approaching an era in which peace treaties will not only be recorded on paper, but will also become inscribed in the hearts of men.

◄ **EARTH.** **On December 7, 1972, just hours after takeoff from the Kennedy Space Center in Florida, the crew of Apollo 17 found themselves aligned with Earth and Sun, enabling them to take this full disk view of Earth. The entire continent of Africa, much of the ice-locked continent of Antarctica, and small portions of Europe and the Asia mainland are visible.**

Science and Religion

During the last century, and part of the one before, it was widely held that there was an unreconcilable conflict between knowledge and belief. The opinion prevailed among advanced minds that it was time that belief should be replaced increasingly by knowledge; belief that did not itself rest on knowledge was superstition, and as such had to be opposed. According to this conception, the sole function of education was to open the way to thinking and knowing, and the school, as the outstanding organ for the people's education, must serve that end exclusively.

One will probably find but rarely, if at all, the rationalistic standpoint expressed in such crass form; for any sensible man would see at once how one-sided is such a statement of the position. But it is just as well to state a thesis starkly and nakedly, if one wants to clear up one's mind as to its nature.

It is true that convictions can best be supported with experience and clear thinking. On this point one must agree unreservedly with the extreme rationalist. The weak point of his conception is, however, this, that those convictions which are necessary and determinant for our conduct and judgments cannot be found solely along this solid scientific way.

◀ **ESO 510 G 13 shows a galaxy that has an unusual twisted disk structure. Gravitational forces distort the structures of the galaxies as their stars, gas, and dust merge together in a process that takes millions of years.**

For the scientific method can teach us nothing else beyond how facts are related to, and conditioned by, each other. The aspiration toward such objective knowledge belongs to the highest of which man is capable, and you will certainly not suspect me of wishing to belittle the achievements and the heroic efforts of man in this sphere. Yet it is equally clear that knowledge of what *is* does not open the door directly to what *should be.* One can have the clearest and most complete knowledge of what *is,* and yet not be able to deduct from that what should be the *goal* of our human aspirations. Objective knowledge provides us with powerful instruments for the achievements of certain ends, but the ultimate goal itself and the longing to reach it must come from another source. And it is hardly necessary to argue for the view that our existence and our activity acquire meaning only by the setting up of such a goal and of corresponding values. The knowledge of truth as such is wonderful, but it is so little capable of acting as a guide that it cannot prove even the justification and the value of the aspiration toward that very knowledge of truth. Here we face, therefore, the limits of the purely rational conception of our existence.

But it must not be assumed that intelligent thinking can play no part in the formation of the goal and of ethical judgments. When someone realizes that for the achievement of an end certain means would be useful, the means itself becomes thereby an end. Intelligence makes clear to us the interrelation of means and ends. But mere thinking cannot give us a sense of the ultimate and fundamental ends. To make clear these fundamental ends and valuations, and to set them fast in the emotional life of the individual, seems to me precisely the most important function which religion has to perform in the social life of man. And if one asks whence derives the authority of such fundamental ends, since they cannot be stated and justified merely by reason, one can only answer: they exist in a healthy society as powerful traditions, which act upon the conduct and aspirations and judgments of the individuals; they are there, that is, as something living, without its being necessary to find justification for their existence. They come into being not through demonstration but through revelation, through the medium of

▲ **THE TADPOLE.** With a long tail of stars and gas streaming from it, the Tadpole galaxy is seen in this stunning image. The tail has been created during a collision with a smaller galaxy, visible in the upper part of the large spiral.

▲ **NGC 5866** is a disk galaxy of type "S 0" (pronounced s-zero). Viewed face on, it would look like a smooth, flat disk with little spiral structure. Here we see it tilted nearly edge on to our line-of-sight. It has a diameter of roughly 60,000 light-years (18,400 parsecs) only two-thirds the diameter of the Milky Way, although its mass is similar to our galaxy.

powerful personalities. One must not attempt to justify them, but rather to sense their nature simply and clearly.

The highest principles for our aspirations and judgments are given to us in the Jewish-Christian religious tradition. It is a very high goal which, with our weak powers, we can reach only very inadequately, but which gives a sure foundation to our aspirations and valuations. If one were to take that goal out of its religious form and look merely at its purely human side, one might state it perhaps thus: free and responsible development of the individual, so that he may place his powers freely and gladly in the service of all mankind.

There is no room in this for the divinization of a nation, of a class, let alone of an individual. Are we not all children of one father, as it is said in religious language? Indeed, even the divinization of humanity, as a totality, would not be in the spirit of that ideal. It is only to the individual that a soul is given. And the high destiny of the individual is to serve rather than to rule, or to impose himself in any other way.

If one holds these high principles clearly before one's eyes, and compares them with the life and spirit of our times, then it appears glaringly that civilized mankind finds itself at present in grave danger. In the totalitarian states it is the rulers themselves who strive actually to destroy that spirit of humanity. In less threatened parts it is nationalism and intolerance, as well as the oppression of the individuals by economic means, which threaten to choke these most precious traditions.

A realization of how great is the danger is spreading, however, among thinking people, and there is much search for means with which to meet the danger—means in the field of national and international politics, of legislation, or organization in general. Such efforts are, no doubt, greatly needed. Yet the ancients knew something which we seem to have forgotten. All means prove but a blunt instrument, if they have not behind them a living spirit. But if the longing for the achievement of the goal is powerfully alive within us, then shall we not lack the strength to find the means for reaching the goal and for translating it into deeds.

. . .

It would not be difficult to come to an agreement as to what we understand by science. Science is the century-old endeavor to bring together by means of systematic thought the perceptible phenomena of this world into a thorough-going an association as possible. To put it boldly, it is the attempt at the posterior reconstruction of existence by the process of conceptualization. But when asking myself what religion is I cannot think of the answer so easily. And even after finding an answer which may satisfy me at this particular moment, I still remain convinced that I can never under any circumstances bring together, even to a slight extent, the thoughts of all those who have given this question serious consideration.

At first, then, instead of asking what religion is I should prefer to ask what characterizes the aspirations of a person who gives me the impression of being religious: a person who is religiously enlightened appears to me to be one who has, to the best of his ability, liberated himself from the fetters of his selfish desires and is preoccupied with thoughts, feelings, and aspirations to which he clings because of their superpersonal value. It seems to me that what is important is the force of this superpersonal content and the depth of the conviction concerning its overpowering meaningfulness, regardless of whether any attempt is made to unite this content with a divine Being, for otherwise it would not be possible to count Buddha and Spinoza as religious personalities. Accordingly, a religious person is devout in the sense that he has no doubt of the significance and loftiness of those superpersonal objects and goals which neither require nor are capable of rational foundation. They exist with the same necessity and matter-of-factness as he himself. In this sense religion is the age-old endeavor of mankind to become clearly and completely conscious of these values and goals and constantly to strengthen and extend their effect. If one conceives of religion and science according to these definitions then a conflict between them appears impossible. For science can only ascertain what *is,* but not what *should be,* and outside of its domain value judgments of all kinds remain necessary. Religion, on the other hand, deals only with evaluations of human thought and action: it cannot justifiably speak of facts and relationships between facts. According to this interpretation the well-known conflicts between religion and science in the past must all be ascribed to a misapprehension of the situation which has been described.

For example, a conflict arises when a religious community insists on the absolute truthfulness of all statements recorded in the Bible. This means an intervention on the part of religion into the sphere of science; this is where the struggle of the Church against the doctrines of Galileo and Darwin belongs. On the other hand, representatives of science have often made an attempt to arrive at fundamental judgments with respect to values and ends on the basis of scientific method, and in this way have set themselves in opposition to religion. These conflicts have all sprung from fatal errors.

Now, even though the realms of religion and science in themselves are clearly marked off from each other, nevertheless there exist between the two strong reciprocal relationships and dependencies. Though religion may be that which determines the goal, it has, nevertheless, learned from science, in the broadest sense, what means will contribute to the attainment of the goals it has set up. But science can only be created by those who are thoroughly imbued with the aspiration toward truth and understanding. This source of feeling, however, springs from the sphere of religion. To this there also belongs the faith in the possibility that the regulations valid for the world of existence are rational, that is, comprehensible to reason. I cannot conceive of a genuine scientist without that profound faith. The situation may be expressed by an image: science without religion is lame, religion without science is blind.

Though I have asserted above that in truth a legitimate conflict between religion and science cannot exist, I must nevertheless qualify this assertion once again on an essential point, with reference to the actual content of historical religions. This qualification has to do with the concept of God. During the youthful period of mankind's spiritual evolution human fantasy created gods in man's own image, who, by the operations of their will were supposed to determine, or at any rate to influence, the phenomenal world. Man sought to alter the disposition of these gods in his own favor by means of magic and prayer. The idea of God in the religions taught at present is a sublimation of that old concept of the gods. Its anthropomorphic character is shown, for instance, by the fact that men appeal to the Divine Being in prayers and plead for the fulfillment of their wishes.

Nobody, certainly, will deny that the idea of the existence of an omnipotent, just, and omnibeneficent personal God is able to accord man solace, help, and guidance; also, by virtue of its simplicity it is accessible to the most undeveloped mind. But, on the other hand, there are decisive weaknesses attached to this idea in itself, which have been painfully felt since the beginning of history. That is, if this being is omnipotent, then every occurrence, including every human action, every human thought, and every human feeling and aspiration is also His work; how is it possible to think of holding men responsible for their deeds and thoughts before such an almighty Being? In giving out punishment and rewards He would to a certain extent be passing judgment on Himself. How can this be combined with the goodness and righteousness ascribed to Him?

The main source of the present-day conflicts between the spheres of religion and of science lies in this concept of a personal God. It is the aim of science to establish general rules which determine the reciprocal connection of objects and events in time and space. For these rules, or laws of nature, absolutely general validity is required—not proven. It is mainly a program, and faith in the possibility of its accomplishment in principle is only founded on partial successes. But hardly anyone could be found who would deny these partial successes and ascribe them to human self-deception. The fact that on the basis of such laws we are able to predict the temporal behavior of phenomena in certain domains with great precision and certainty is deeply embedded in the consciousness of the modern man, even though he may have grasped very little of the contents of those laws. He need only consider that planetary courses within the solar system may be calculated in advance with great exactitude on the basis of a limited number of simple laws. In a similar way, though not with the same precision, it is possible to calculate in advance the mode of operation of an electric motor,

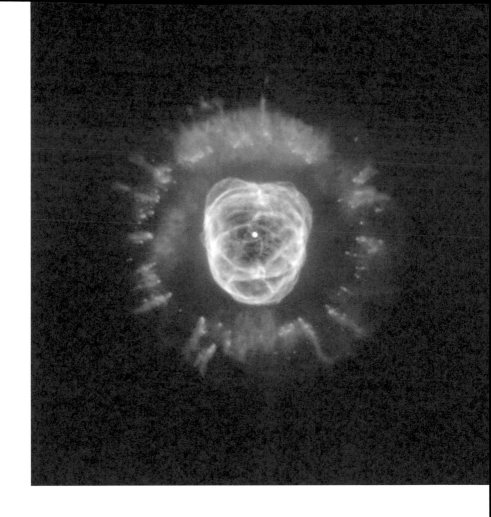

▲ **THE ESKIMO NEBULA.** The majestic planetary nebula
NGC 2392 is the glowing remains of a dying, Sun-like star.
This stellar relic, first spied by William Herschel in 1787,
is nicknamed the "Eskimo" nebula because, when viewed
through ground-based telescopes, it resembles a face
surrounded by a fur parka. The "parka" is really a disk of
material embellished with a ring of comet-shaped objects,
with their tails streaming away from the central, dying star.
The Eskimo's "face" also contains some fascinating details.
Although this bright central region resembles a ball of twine,
it is, in reality, a bubble of material being blown into space by
the central stars intense "wind" of high-speed material.

▲ **PELICAN NEBULA IONIZATION FRONT.** This image
reveals many previously unseen shock waves, evidence
for powerful outflows from newly formed stars embedded
within the molecular clouds that rim the nebula.

a transmission system, or of a wireless apparatus, even when dealing with a novel development.

To be sure, when the number of factors coming into play in a phenomenological complex is too large, scientific method in most cases fails us. One need only think of the weather, in which case prediction even for a few days ahead is impossible. Nevertheless no one doubts that we are confronted with a causal connection whose causal components are in the main known to us. Occurrences in this domain are beyond the reach of exact prediction because of the variety of factors in operation, not because of any lack of order in nature.

We have penetrated far less deeply into the regularities obtaining within the realm of living things, but deeply enough nevertheless to sense at least the rule of fixed necessity. One need only think of the systematic order in heredity, and in the effect of poisons, as for instance alcohol, on the behavior of organic beings. What is still lacking here is a grasp of connections of profound generality, but not a knowledge of order in itself.

The more a man is imbued with the ordered regularity of all events the firmer becomes his conviction that there is no room left by the side of this ordered regularity for causes of a different nature. For him neither the rule of human nor the rule of divine will exists as an independent cause of natural events. To be sure, the doctrine of a personal God interfering with natural events could never be *refuted*, in the real sense, by science, for this doctrine can always take refuge in those domains in which scientific knowledge has not yet been able to set foot.

But I am persuaded that such behavior on the part of the representatives of religion would not only be unworthy but also fatal. For a doctrine which is able to maintain itself not in clear light but only in the dark, will of necessity lose its effect on mankind, with incalculable harm to human progress. In their struggle for the ethical good, teachers of religion must have the stature to give up the doctrine of a personal God, that is, give up that source of fear and hope which in the past placed such vast power in the hands of priests. In their labors they will have to avail themselves of those forces which are capable of cultivating the Good, the True, and the Beautiful in humanity itself. This is, to be sure, a more difficult but an incomparably more worthy task. (This thought is convincingly presented in Herbert Samuel's book, *Belief and Action*.) After religious teachers accomplish the refining process indicated they will surely recognize with joy that true religion has been ennobled and made more profound by scientific knowledge.

If it is one of the goals of religion to liberate mankind as far as possible from the bondage of egocentric cravings, desires, and fears, scientific reasoning can aid religion in yet another sense. Although it is true that it is the goal of science to discover rules which permit the association and foretelling of facts, this is not its only aim. It also seeks to reduce the connections discovered to the smallest possible number of mutually independent conceptual elements. It is in this striving after the rational unification of the manifold that it encounters its greatest successes, even though it is precisely this attempt which causes it to run the greatest risk of falling a prey to illusions. But whoever has undergone the intense experience of successful advances made in this domain is moved by profound reverence for the rationality made manifest in existence. By way of the understanding he achieves a far-reaching emancipation from the shackles of personal hopes and desires, and thereby attains that humble attitude of mind toward the grandeur of reason incarnate in existence, and which, in its profoundest depths, is inaccessible to man. This attitude, however, appears to me to be religious, in the highest sense of the word. And so it seems to me that science not only purifies the religious impulse of the dross of its anthropomorphism but also contributes to a religious spiritualization of our understanding of life.

The further the spiritual evolution of mankind advances, the more certain it seems to me that the path to genuine religiosity does not lie through the fear of life, and the fear of death, and blind faith, but through striving after rational knowledge. In this sense I believe that the priest must become a teacher if he wishes to do justice to his lofty educational mission.

▲ **LIGHT AND SHADOW IN THE CARINA
NEBULA** (NGC 3372). When 19th century
astronomer Sir John Herschel spied a
swirling cloud of gas with a hole punched
through it, he dubbed it the Keyhole Nebula.
Now the Hubble telescope has taken a peek
at this region, and the resulting image reveals
previously unseen details of the Keyhole's
mysterious, complex structure.

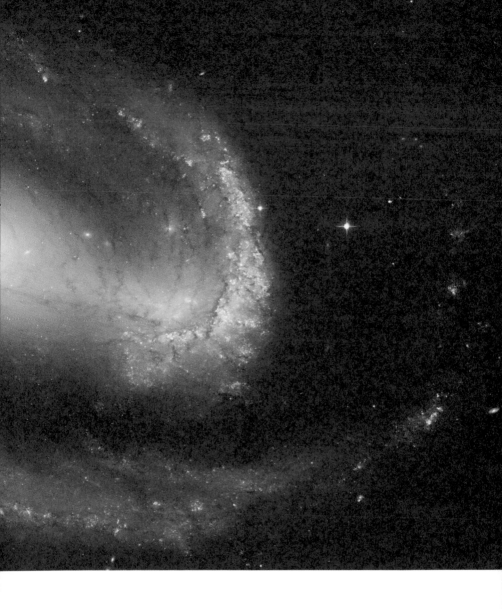

▲ **NGC 1300** is considered to be prototypical of barred spiral galaxies. Barred spirals differ from normal spiral galaxies in that the arms of the galaxy do not spiral all the way into the center, but are connected to the two ends of a straight bar of stars containing the nucleus at its center. The nucleus of NGC 1300 shows an extraordinary "grand-design" spiral structure that is about 3,300 light-years (1 kiloparsec) in diameter.

Science and Society

There are two ways in which science affects human affairs. The first is familiar to everyone: Directly, and to an even greater extent indirectly, science produces aids that have completely transformed human existence. The second way is educational in character—it works on the mind. Although it may appear less obvious to cursory examination, it is no less incisive than the first.

The most conspicuous practical effect of science is that it makes possible the contriving of things that enrich life, though they complicate it at the same time—inventions such as the steam engine, the railway, electric power and light, the telegraph, radio, automobile, airplane, dynamite, etc. To these must be added the life-preserving achievements of biology and medicine, especially the production of pain relievers and preservative methods of storing food. The greatest practical benefit which all these inventions confer on man I see in the fact that they liberate him from the excessive muscular drudgery that was once indispensable for the preservation of bare existence.

◄ **SOLAR SYSTEM MONTAGE.** These planetary images were taken by spacecraft managed by the Jet Propulsion Laboratory in Pasadena, California. Included are (from top to bottom) images of Mercury, Venus, Earth (and Moon), Mars, Jupiter, Saturn, Uranus, and Neptune. Pluto (recently demoted to Keiper Belt object) is not shown as no spacecraft has yet visited it.

Insofar as we may at all claim that slavery has been abolished today, we owe its abolition to the practical consequences of science.

On the other hand, technology—or applied science—has confronted mankind with problems of profound gravity. The very survival of mankind depends on a satisfactory solution of these problems. It is a matter of creating the kind of social institutions and traditions without which the new tools must inevitably bring disaster of the worst kind.

Mechanical means of production in an unorganized economy have had the result that a substantial proportion of mankind is no longer needed for the production of goods and is thus excluded from the process of economic circulation. The immediate consequences are the weakening of purchasing power and the devaluation of labor because of excessive competition, and these give rise, at ever shortening intervals, to a grave paralysis in the production of goods. Ownership of the means of production, on the other hand, carries a power to which the traditional safeguards of our political institutions are unequal. Mankind is caught up in a struggle for adaptation to these new conditions—a struggle that may bring true liberation, if our generation shows itself equal to the task.

Technology has also shortened distances and created new and extraordinarily effective means of destruction which, in the hands of nations claiming unrestricted freedom of action, become threats to the security and very survival of mankind. This situation requires a single judicial and executive power for the entire planet, and the creation of such a central authority is desperately opposed by national traditions. Here too we are in the midst of a struggle whose issue will decide the fate of all of us.

▶ **JUPITER AND ITS VOLCANIC MOON, IO.** This montage of Jupiter and its volcanic moon Io was taken during the New Horizon spacecraft's Jupiter flyby in early 2007. Blue areas denote high-altitude clouds and hazes, and red indicates deeper clouds. The prominent bluish-white oval is the Great Red Spot.

▲ **MERCURY—IN COLOR.** The Messenger
spacecraft transmitted to Earth the first high-
resolution image of Mercury by a spacecraft in
over 30 years. The Messenger data allow Mercury
to be seen in a variety of high-resolution color
views not previously possible.

Means of communication, finally—reproduction processes for the printed word, and the radio, when combined with modern weapons have made it possible to place body and soul under bondage to a central authority—and here is a third source of danger to mankind. Modern tyrannies and their destructive effects show plainly how far we are from exploiting these achievements organizationally for the benefit of mankind. Here too circumstances require an international solution, with the psychological foundation for such a solution not yet laid.

Let us now turn to the intellectual effects that proceed from science. In prescientific times it was not possible by means of thought alone to attain results that all mankind could have accepted as certain and necessary. Still less was there a conviction that all that happens in nature is subject to inexorable laws. The fragmentary character of natural law, as seen by the primitive observer, was such as to foster a belief in ghosts and spirits. Hence even today primitive man lives in constant fear that supernatural and arbitrary forces will intervene in his destiny.

It stands to the everlasting credit of science that by acting on the human mind it has overcome man's insecurity before himself and before nature. In creating elementary mathematics the Greeks for the first time wrought a system of thought whose conclusions no one could escape. The scientists of the Renaissance then devised the combination of systematic experiment with mathematical method. This union made possible such precision in the formulation of natural laws and such certainty in checking them by experience that as a result there was no longer room for basic differences of opinion in natural science. Since that time each generation has built up the heritage of knowledge and understanding, without the slightest danger of a crisis that might jeopardize the whole structure.

The general public may be able to follow the details of scientific research to only a modest degree; but it can register at least one great and important gain: confidence that human thought is dependable and natural law universal.

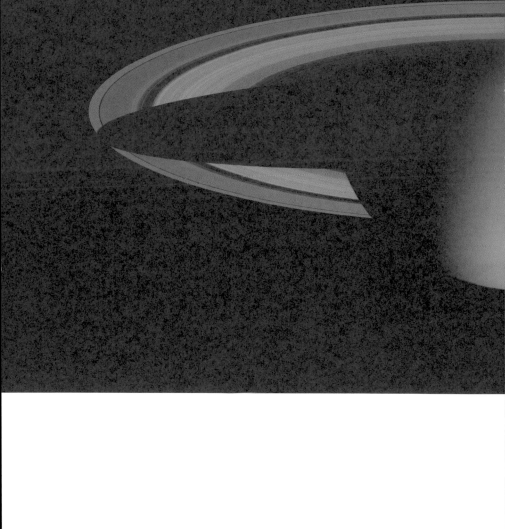

▲ **THE GREATEST SATURN PORTRAIT... Yet.** While cruising around Saturn in early October 2004 the Cassini spacecraft captured a series of images that have been composed into the largest, most detailed, global natural color view of Saturn and its rings ever made. This grand mosaic consists of 126 images acquired in a tile-like fashion, covering one end of Saturn's rings to the other and the entire planet in between.

Art and Creativity

Personally, I experience the greatest degree of pleasure in having contact with works of art. They furnish me with happy feelings of an intensity that I cannot derive from other sources.

. . .

True art is characterized by an irresistible urge in the creative artist.

. . .

The greatest scientists are artists as well.

. . .

Where the world ceases to be the scene of our personal hopes and wishes, where we face it as free beings, admiring, questioning, and observing, there we enter the realm of art and science. We do science when we reconstruct in the language of logic what we have seen and experienced; we do art when we communicate through forms whose connections are not accessible to the conscious mind yet we intuitively recognize them as meaningful.

. . .

◄ **HEAVY-WEIGHT STARS LIGHT UP NEBULA**
NGC 6357: Pismis 24, NGC 6357, CI Pismis.

The development of science and creative activities requires freedom, independence of thought from the restrictions of authoritarian and social prejudices.

.　　.　　.

We cannot solve our problems with the same thinking we used when we created them.

▼ **M 17, also known as the Omega or Swan nebula, is located about 5500 light-years away in the constellation Sagittarius. Wave-like patterns of gas have been sculpted and illuminated by a torrent of ultraviolet radiation from young, massive stars, which lie outside the picture to the upper left. Intense heat and pressure cause some material to stream away, creating the glowing veil of greenish gas that masks background structures.**

▲ **M 82, NGC 3034.** Composite of multi-wavelength images of the active galaxy M 82 from the three great observatories: Hubble space telescope, Chandra X-Ray Observatory, and Spitzer space telescope. X-ray data appears here in blue; infrared light appears in red; observations of hydrogen emission appear in orange, and the bluest visible light appears in yellow-green.

▲ **HUBBLE EXOPLANET SEARCH FIELD IN SAGITTARIUS.** NASA's Hubble
space telescope has discovered 16 extrasolar planet candidates orbiting
a variety of distant stars in the central region of our Milky Way galaxy.
Hubble looked for extrasolar planets farther than has ever successfully been
searched. It peered at 180,000 stars in the crowded central bulge of our

galaxy 26,000 light-years away. That is one-quarter the diameter of the
Milky Way's spiral disk. Five of the newly discovered planets represent a
new extreme type of planet not found in any nearby searches. Dubbed
ultra-short-period planets (USPPs), these worlds whirl around their stars
in less than one Earth day.

▲ **ABSTRACT ART FOUND IN
THE ORION NEBULA, M 42, NGC 1976.**
Treasures reside within the nearby,
intense star-forming region called
the Orion nebula. This Hubble space
telescope color mosaic shows deeply
contrasting areas of light and dark that
blends with a palette of colors mix to
form rich swirls and fluid motions. It is
enough to make even the best artists
stand back and admire the work.

▲ **SPITZER AND HUBBLE CREATE COLORFUL MASTERPIECE.**
This beautiful image from the Spitzer and Hubble space telescopes looks
more like an abstract painting than a cosmic snapshot. The magnificent
masterpiece shows the Orion nebula in an explosion of infrared, ultraviolet
and visible-light colors. It was "painted" by hundreds of baby stars on a
canvas of gas and dust, with intense ultraviolet light and strong stellar
winds as brushes. At the heart of the artwork is a set of four monstrously
massive stars, collectively called Trapezium. These behemoths are
approximately 100,000 times brighter than our Sun.

Imagination

Imagination is more important than knowledge. For knowledge is limited, whereas imagination embraces the entire world, stimulating progress, giving birth to evolution.

◄ **HORSEHEAD NEBULA, NGC 2024 AND IC 434.**
The distinctive red emission nebula IC 434 is a result of radiation from Sigma Orionis interacting with the surface of a dusty cloud of gas. From it projects the dark shape of the head of a horse. The brightest star, Zeta Orionis, is easily visible to the unaided eye as the easternmost star in the line of three which form Orion's belt.

▲ **EAGLE NEBULA, M 16, NGC 6611, IC 4703.** Resembling a
fanciful fairy-tale creature, this object is actually a billowing mass of
cold gas and dust coming from a stellar nursery called the Eagle nebula.
It measures 9.5 light-years or about 92 trillion km (57 trillion miles)
long, about twice the distance from our Sun to the next nearest star.
Stars in the Eagle nebula are born in clouds of cold hydrogen gas that
reside in chaotic neighborhoods, where energy from young stars

sculpts fantasy-like landscapes in the gas. An incubator for newborn stars, such enormous clouds breed stars by dense gas collapsing under gravity. Other stars may be formed due to pressure from gas that has been heated by the neighboring hot stars. The bumps and fingers of material in the center are examples of these stellar birthing areas. These regions may look small but they are roughly the size of our solar system.

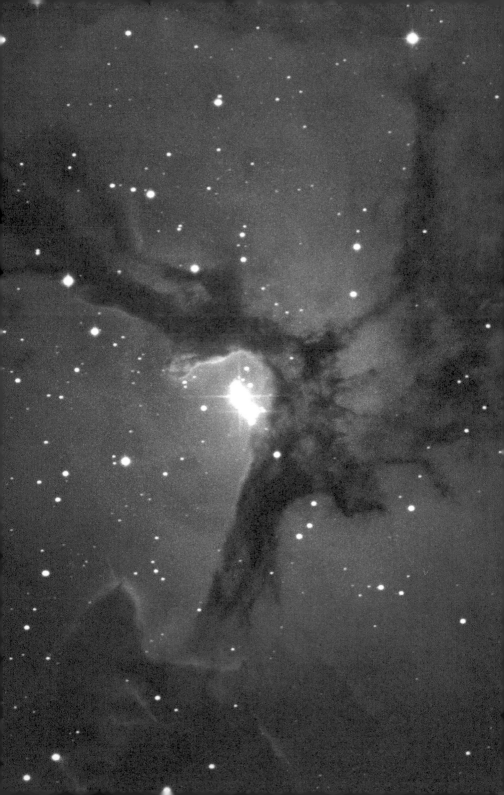

Curiosity

The important thing is not to stop questioning. Curiosity has its own reason for existence. One cannot help but be in awe when [one] contemplates the mysteries of eternity, of life, of the marvelous structure of reality. It is enough if one tries merely to comprehend a little of this mystery each day. Never lose a holy curiosity.

.　　.　　.

It is nothing short of a miracle that modern methods of instruction have not yet entirely strangled the holy curiosity of inquiry.

◀ **STARS IN THE TRIFID NEBULA.** Distributed throughout the Milky Way are vast clouds of hydrogen mixed with tiny dust grains. Hydrogen is only visible when it is illuminated by very hot stars. Light from these stars is sufficiently rich in ultraviolet light to cause the gas to glow with its characteristic red color.

Nature

Look deep, deep into nature, and then you will understand everything better.

.   .   .

Nature hides her secrets through her intrinsic grandeur, not through deception.

◄ **DEEPEST VIEW EVER OF THE UNIVERSE.**
The deepest portrait of the visible universe ever
achieved by humankind. Called Hubble Ultra Deep Field
(HUDF), a million-second-long exposure reveals the first
galaxies to emerge from the so-called "dark ages",
time shortly after the big bang when the first stars
reheated a cold, dark universe. The new image should
offer insights into what types of objects reheated
our universe long ago.

Eternal Mystery

The eternal mystery of the world is its comprehensibility . . . . That the world is comprehensible is a miracle.

◀ **REFLECTION NEBULA IN ORION, NGC 1973–75–77 is a group of nebulous stars just half a degree north of the much brighter Orion nebula. To the unaided eye, this group appears as a single star, the northernmost "star" in the sword of Orion. The blue nebulosity is mostly starlight scattered by dust, while the distinctive red glow results from stars sufficiently hot to excite wisps of hydrogen. Together, the stars and nebula seen here create an impression of quiet beauty.**

▲ **HODGE 301 IN THE TARANTULA NEBULA**
is a cluster of brilliant, massive stars, near the
edge of the most active starburst region in the
local universe. Located within one of our nearest
galactic neighbors, the Large Magellanic Cloud,
Hodge 301 contains many old stars that have
exploded as supernovae. These stellar fireworks
have blasted material into the surrounding
region at high speeds. As the ejecta plow into
the surrounding Tarantula nebula, they shock
and compress the gas into a multitude of sheets
and filaments.

The Goal of Human Existence

Our age is proud of the progress it has made in man's intellectual development. The search and striving for truth and knowledge is one of the highest
of man's qualities though often the pride is most loudly voiced by those who
strive the least. And certainly we should take care not to make the intellect
our god; it has, of course, powerful muscles, but no personality. It cannot
lead, it can only serve; and it is not fastidious in its choice of a leader. This
characteristic is reflected in the qualities of its priests, the intellectuals. The
intellect has a sharp eye for methods and tools, but is blind to ends and values. So it is no wonder that this fatal blindness is handed on from old to
young and today involves a whole generation.

Our Jewish forbears, the prophets and the old Chinese sages understood and proclaimed that the most important factor in giving shape to our
human existence is the setting up and establishment of a goal; the goal
being a community of free and happy human beings who by constant inward
endeavor strive to liberate themselves from the inheritance of anti-social and
destructive instincts. In this effort the intellect can be the most powerful aid.
The fruits of intellectual effort, together with the striving itself, in cooperation
with the creative activity of the artist, lend content and meaning to life.

But today the rude passions of man reign in our world, more unrestrained
than ever before…

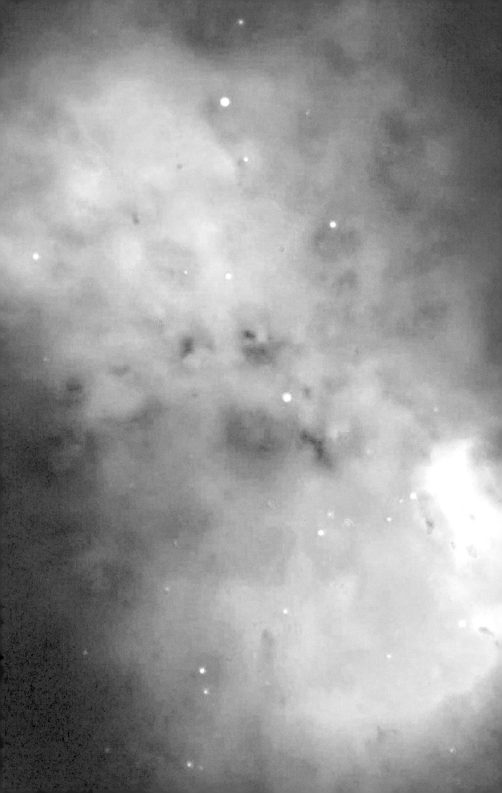

My Credo

Our situation on this earth seems strange. Every one of us appears here involuntarily and uninvited for a short stay, without knowing the whys and the wherefore. In our daily lives we only feel that man is here for the sake of others, for those whom we love and for many other beings whose fate is connected with our own.

I am often worried at the thought that my life is based to such a large extent on the work of my fellow human beings and I am aware of my great indebtedness to them.

I do not believe in freedom of the will. Schopenhauer's words: "Man can do what he wants, but he cannot will what he wills" accompany me in all situations throughout my life and reconcile me with the actions of others even if they are rather painful to me. This awareness of the lack of freedom of will preserves me from taking too seriously myself and my fellow men as acting and deciding individuals and from losing my temper.

◄ **THE DUMBBELL NEBULA** — also known as **Messier 27 or NGC 6853** — was first described by the French astronomer and comet hunter Charles Messier who found it in 1764 and included it as no. 27 in his famous list of extended sky objects. It consists of very rarified gas that has been ejected from the hot central star (well visible on this photo), now in one of the last evolutionary stages.

▲ **SMALL PORTION OF CYGNUS LOOP NEBULA.** This nebula is an expanding blast wave from a stellar cataclysm, a supernova explosion, which occurred about 15,000 years ago. The supernova blast wave, which is moving from left to right across the picture, has recently hit a cloud of denser-than-average interstellar gas. This collision drives shock waves into the cloud that heats interstellar gas, causing it to glow.

I never coveted affluence and luxury and even despise them a good deal. My passion for social justice has often brought me into conflict with people, as did my aversion to any obligation and dependence I do not regard as absolutely necessary. I always have a high regard for the individual and have an insuperable distaste for violence and clubmanship. All these motives made me into a passionate pacifist and anti-militarist. I am against any nationalism, even in the guise of mere patriotism.

Privileges based on position and property have always seemed to me unjust and pernicious, as did any exaggerated personality cult. I am an adherent of the ideal of democracy, although I well know the weaknesses of the democratic form of government. Social equality and economic protection of the individual appeared to me always as the important communal aims of the state.

Although I am a typical loner in daily life, my consciousness of belonging to the invisible community of those who strive for truth, beauty, and justice has preserved me from feeling isolated.

The most beautiful and deepest experience a man can have is the sense of the mysterious. It is the underlying principle of religion as well as all serious endeavour in art and science. He who never had this experience seems to me, if not dead, then at least blind. To sense that behind anything that can be experienced there is a something that our mind cannot grasp and whose beauty and sublimity reaches us only indirectly and as a feeble reflection, this is religiousness. In this sense I am religious. To me it suffices to wonder at these secrets and to attempt humbly to grasp with my mind a mere image of the lofty structure of all that there is.

The Religious Spirit of Science

You will hardly find one among the profounder sort of scientific minds without a peculiar religious feeling of his own. But it is different from the religion of the naive man. For the latter, God is a being from whose care one hopes to benefit and whose punishment one fears; a sublimation of a feeling similar to that of a child for its father, a being to whom one stands to some extent in a personal relation, however deeply it may be tinged with awe.

But the scientist is possessed by the sense of universal causation. The future, to him, is every whit as necessary and determined as the past. There is nothing divine about morality; it is a purely human affair. His religious feeling takes the form of a rapturous amazement at the harmony of natural law, which reveals an intelligence of such superiority that, compared with it, all the systematic thinking and acting of human beings is an utterly insignificant reflection. This feeling is the guiding principle of his life and work, in so far as he succeeds in keeping himself from the shackles of selfish desire. It is beyond question closely akin to that which has possessed the religious geniuses of all ages.

◀ **ANTARES AND THE RHO OPHIUCHI DARK CLOUD.** Between Ophiuchi and Scorpius lies a dusty region that contains some of the most colorful and spectacular nebulae known. The upper part of the picture is filled with the bluish glow of light from hot stars reflected by a huge, cool cloud of dust and gas where such stars are born. The red supergiant Antares (600 light years away) dominates the lower half of this cosmic landscape. Sigma Scorpii (735 light years distant) at the right of the picture is a red emission nebula, completing the most comprehensive collection of nebular types ever seen in one photograph.

▲ **THE GREAT NEBULA IN ORION** The nearby
stellar nursery known as the Orion nebula excites the
imagination like few astronomical sights. The nebula's
glowing gas surrounds hot young stars at the edge of
an immense interstellar molecular cloud, about 1,500
light-years away. The whole Orion nebula cloud complex,
which includes the Horsehead nebula, will slowly
disperse over the next 100,000 years.

Afterword

We Have Survived Fifty Years

Fundamental questions have confronted humanity for thousands of years. What is the nature of the world and how can humanity understand the world? Scientists, theologians, philosophers, and laypeople have considered these mysteries and developed different views of the world. Worldviews emerge and change with the introduction of new scientific knowledge and insights and the passage of historical events and human achievements over time.

Einstein's work radically transformed our understanding of space, time, matter and energy. His theories provide some answers to fundamental questions about the structure of the planet, most of which contain invisible elements and abstractions, far removed from everyday experience. The history of human experience and culture has always been influenced by the invisible forces of nature, although we may know very little about these forces.

More than ever, scientific questions are also questions for society at large. Which technologies should be our basis for generating energy? How can we prevent the development of new nuclear weapons and their proliferation? Should scientific information be freely available on the internet? Should we prevent the dissemination of potentially dangerous information? Will Einstein's legacy also include the demand for a responsible, ethical, free, and democratic culture of knowledge?

After the Second World War, a new public awareness of scientific responsibility has developed against the background of Nazi war crimes and the development of the atomic bombs that destroyed Hiroshima and Nagasaki. In the twenty-first century, science faces political challenges that reflect the industrial dimensions of the scientific system. Large-scale research projects to explore outer space are expensive and divert limited resources away from social needs. Biotechnology and stem-cell research involve both opportunities and risks. Where should we draw the limits of scientific freedom? And who should make these decisions?

In open societies, it is necessary to mediate between scientific freedom and the need to democratically justify publicly financed science. Public awareness of the value and nature of science and a willingness to be appropriately informed about controversial issues are prerequisites for a meaningful discussion about science.

The unleashed power of the atom bomb has changed everything except our modes of thinking... the solution to this problem lies in the heart of mankind. *(1946)*

In the heart of mankind? What is in our hearts? Einstein believed we do know what is good and what is bad. We do what is good because it is good, not expecting a reward from a God. We do not have to learn to be good, but we must learn a new way of thinking, if we are to avoid destroying ourselves and our world. We must learn that we have a common enemy: genocidal nuclear weapons. We must abolish that enemy before it abolishes us.

International Physicians for the Prevention of Nuclear War was founded in 1980, with the belief that, if people really understood the nature of nuclear war, they would demand of the governments of the nuclear-weapon states to abolish nuclear weapons, and governments would heed such a demand. We succeeded in making people understand the nature of the nuclear threat, for the people demanded and still demand the abolition of nuclear weapons. But the governments have not complied. People become tired and disillusioned when their entreaties are not heard. The governments have failed and democracy has failed.

Do the people despair? No, they will find other ways to change the world.

Almost every great sustained advance in the world since 1945 has been brought about by ordinary people working together without violence. The liberation of India and other colonized countries, the democratization of South America and, most remarkably and unexpectedly, the fall of the Soviet Empire are the results of non-violent action. Every gain for democracy and freedom has been won by the voluntary cooperation of people, mostly by non-violent means. Wars have been fought in Vietnam, the Congo, Rwanda, Sudan, Sri Lanka, Iraq, and scores of other places and the result has been the death of millions, but democracy and freedom live on. War solves no problem.

Today there exists in the world a feeling of our common destiny, our common responsibility, which never existed before. We have seen the image of our fragile blue planet rising over the barren surface of the moon, life over death. We view the Earth with tenderness as we would a child, our responsibility. Our planet, our responsibility, tenderness; a new relationship between man and Earth; a new feeling of us and we. Young people of the world now communicate by internet, share their thoughts, their music, their concerns. "We are the children of the world."

> I hold that mankind is approaching an era in which peace treaties will not only be recorded on paper, but will also become inscribed in the hearts of men. *(1946)*

A new way of thinking? Yes, it is underway. There are temporary setbacks. The biggest demonstrations ever seen in the world, against the attack on Iraq, did not stop the war. The doctrine of nuclear deterrence has been expanded into a new military doctrine that declares nuclear weapons to be "usable" in war-fighting. This new doctrine has sneaked past the unwary general public and caused a fraction of the outrage it would have if its implications had been known and understood.

There exist in the world two superpowers. One is the mightiest military power in human history. It believes it can rule the world with its instruments of death. The other superpower is civil society. This is a new situation. Ordinary people know that another world is possible, and that it is their right and their responsibility to defend humanity against militarism and create a new world order of peace and justice.

▲ **Albert Einstein's 70th birthday party, 1949.**

We cannot predict the outcome of this struggle. The mighty control not only the instruments of death but also, to a large extent, the images that reach us through the media. These powerful images create a feeling of them instead of us, of hate instead of compassion, of despair instead of hope. Will the new cellular phone-camera-internet connectivity change our perspective? Will we see the soldiers and tanks through the eyes of the victims instead of the "embedded journalist"?

A new way of thinking? The enemy is not the peoples of other countries, not the Earth that fights back when we destroy its life-support systems. The enemy is war and militarism, "our" terrorists as much as "their" terrorism, nuclear weapons, environmental destruction, poverty, exploitation, unnecessary deaths of 15,000 infants every day.

We must inoculate our children against militarism,
by educating them in the spirit of pacifism....
Our schoolbooks glorify war and conceal its
horrors. They indoctrinate children with hatred.
I would teach peace rather than war,
love rather than hate. *(1979)*

What would Einstein think of the world today? He would be happy, maybe a bit surprised, that we are still here, not exterminated by nuclear war, not yet. He would be unhappy that militarism and trust in weapons are as strong as ever, although the Cold War is over. He would find it incomprehensible that religious fundamentalism and intolerance are becoming more influential in many countries, even in highly developed democracies. Why do we promote conflict between civilizations instead of dialogue?

I should have become a watchmaker *(1965)*

No, we see Einstein smiling when he said this. He did not regret that he had dedicated his life to discovering the laws of the universe, to reading the thoughts of God. We are grateful that he also shared with us his lifelong thoughts on what it is to be human. Einstein's global view is shared by more and more people. The responsibility rests with us to honor and carry out his legacy.

We cannot but be inspired when we read Einstein's quotes and meditate over the images of the strange beauty of the universe which is captured in this book.

Ron McCoy
Kuala Lumpur, Malaysia

Gunnar Westberg
Goteborg, Sweden

Dr. McCoy and Dr. Westberg are former co-presidents of International Physicians for the Prevention of Nuclear War. IPPNW was the recipient of the Nobel Peace Prize in 1985.

Peace Resources

The Albert Einstein Institution (AEI)
PO Box 455
East Boston, MA 02139
www.aeinstein.org

Cat Lovers Against the Bomb (CLAB-NR)
PO Box 83466
Lincoln, NE 68501
www.catloversagainstthebomb.org

Center for Advanced Military Science (CAMS)
Institute of Science, Technology and Public Policy
1000 North 4th Street
Fairfield, IA 52557
www.istpp.org

Code Pink
1247 E Street SE
Washington, DC 20003
www.codepink4peace.org

David Lynch Foundation
Operation Warrior Wellness
654 Madison Avenue Suite 806
New York, NY 10065
www.DavidLynchFoundation.org

Defend Science
2124 Kittridge Street 182
Berkeley, CA 94704
www.defendscience.org

DC Poets Against the War
626 Allison Street NW
Washington, DC 20011
www.dcpaw.poetrymutual.org

Freedom Writers Foundation
PO Box 41505
Long Beach, CA 90853
www.freedomwritersfoundation.org

Global Union of Scientists for Peace
2000 Capital Boulevard
Fairfield, IA 52556
www.gusp.org

Grandmothers for Peace International
PO Box 1292
Elk Grove, CA 95759
www.grandmothersforpeace.org

The International Committee of Artists for Peace (ICAP)
201 Ocean Ave. Ste. 1708P
Santa Monica, CA 90402
www.icapeace.org

International Fellowship of Reconciliation (IFOR)
Spoorstraat 38, 1815 BK
Alkmaar, The Netherlands
www.ifor-mir.org

International Peace Bureau (IPB)
41 Rue de Zurich, CH 1201
Geneva, Switzerland
www.ipb.org

International Peace Prize Trondheim
The Student Peace Prize
www.ISFiT.org
www.thestudentpeaceprize.org

International Physicians for Prevention of Nuclear War
IPPNW Central Office
66-70 Union Square 204
Somerville, MA 02143
www.ippnw.org

Journal of Peace Research
International Peace Research Institute
PO Box 9229 Grønland
NO 0134 Oslo, Norway
www.prio.no

Nobel Women's Initiative
1 Nicholas Street Ste. 430
Ottawa, ON KIN 7B7, Canada
www.nobelwomensinitiative.org

Peace Magazine
PO Box 248, Toronto P
Toronto ON M5S 2S7, Canada
www.peacemagazine.org

Peace4Kids
PO Box 5347
Compton, CA 90224
www.Peace4Kids.org

Pennies for Peace
c/o Central Asia Institute
PO Box 7209
Bozeman, MT 59771
www.penniesforpeace.org

Physicians for Social Responsibility (PSR)
1111 14th Street NW Ste. 700
Washington, DC 20005
www.psr.org

Playing for Change Foundation
3110 Main St., The Annex
Santa Monica, CA 90405
www.playingforchange.org

Play for Peace
500 North Michigan Ave. Ste. 300
Chicago, IL 60611
www.playforpeace.org

The Pluralism Project
Harvard University
2 Arrow Street, 4th Fl.
Cambridge, MA 02138
www.pluralism.org

Pugwash Conferences on Science and World Affairs
1111 19th Street NW Ste.1200
Washington, DC 20036
www.pugwash.org

Religions for Peace International
777 United Nations Plaza, 9th Fl.
New York, NY 10017
www.religionsforpeaceinternational.org

Resources for Human Development, California
1673 E. 108 St.
Los Angeles, CA 90002
www.rhdca.org

United States Peace Government
2000 Capital Boulevard
Maharishi Vedic City, IA 52556
www.uspeacegovernment.org

University for Peace
Campus Rodrigo Carazo
El Rodeo de Mora,
San José, Costa Rica
PO Box 138 6100
www.upeace.org

Veterans For Peace
216 South Meramec Avenue
St. Louis, MO 63105
www.veteransforpeace.org

Women's Action for New Directions (WAND)
691 Massachusetts Ave.
Arlington, MA 02476
www.wand.org

Women's International League for Peace and Freedom (WILPF)
1 Rue de Varembé,
Case Postale 28 1211
CH Geneva 20, Switzerland
www.wilpfinternational.org

War Resisters' International (WRI)
5 Caledonian Road
London, N1 9DX, UK
www.wri-irg.org phone

Youth Action for Peace (YAP)
Avenue du Parc Royal 3
Brussels, Belgium 1020
www.yap.org

Acknowledgments

The editors of this book thank Alice Calaprice for finding the quote about imagination, searching for picture sources, other important suggestions, and for writing the preface. We also thank Gunnar Westberg and Ron McCoy, former co-presidents of International Physicians for Prevention of Nuclear War for their epilog, David Malin, Craig Peterson, and Neil deGrasse Tyson for valuable suggestions, and the late James Van Allen for his encouragement and for writing the foreword. We thank all those who took time to read and endorse this book. We also sincerely thank Greg Slater, Elizabeth Voelk, Kathleen Powell, Alisdair Davey, Connie Moore, David McKenzie, Mike Gentry, Mary Ann Hager, Christopher Stark, and Xaviant Ford for finding some of the beautiful and inspiring pictures in this book. We also thank all scientists, technicians, as well as professional and amateur astronomers who provided stunning astronomical images and fascinating scientific facts. Special thanks to Carl Johnson for his enthusiastic encouragement. We are very grateful to Gita Brady for her editorial suggestions and limitless compassion, the Albert Schweitzer Fellowship—in particular, Lachlan Forrow and Ian Stevenson—for a computer capable of handling high-resolution photographs, the support of our families and friends, including our literary agent Kristina Holmes, Jim Rubis (former head librarian at the Fairfield Public Library, as well as its helpful staff then and today), Maharishi University of Management Library, and our dear friends David Blair and Jim Bates, fellow lovers of the planets and stars.

Our deepest love and gratitude to His Holiness Maharishi Mahesh Yogi, founder of the Science of Creative Intelligence (SCI), Transcendental Meditation (TM) and TM-Sidhi programs and for profound insights into Vedic knowledge, and to John Hagelin, into pure transcendental consciousness and the Unified Field.

We thank the following for their kind permission to use copyrighted quotes and articles by Einstein in this collection: Bantam Books for excerpts from *The Universe and Dr. Einstein* by Lincoln Barnett, copyright 1966 by Lincoln Barnett, distributed by Morrow & Company; Branden Press, Inc. for excerpts from *Einstein and the Poet: In Search of the Cosmic Man* by William

Hermanns, copyright 1983 by Branden Press, Inc.; Bonanza Books for "The World as I See It," "On Wealth," copyright 1954 by Crown Publishers, Inc., in *Ideas and Opinions;* Doubleday & Company for excerpts from *The Drama of Albert Einstein* by Antonina Vallentin, copyright 1954 by Antonina Vallentin; Dutton Books for excerpts from *Einstein: A Life in Science* by Michael White and John Gribbin, copyright 1993 by Michael White and John Gribbin; Harvard University Press for excerpts from *Einstein: A Centenary Volume* by Anthony P. French, copyright 1979 by The International Commission on Physics Education; Open Court Publishing Company for excerpts from *Albert Einstein: Autobiographical Notes,* copyright 1979 by The Library of Living Philosophers, Inc. and the Albert Einstein Estate; Philosophical Library for "The Meaning of Life," "Good and Evil," "Religion and Science," "The Religiousness of Science," "Christianity and Judaism," Wisdom Library, distributed by Citadel Press, in *The World As I See It;* and "Self-Portrait," "Moral Decay," "Morals and Emotions," "Science and Religion (Part 1 and 2)," "Science and Society," "The Goal of Human Existence," copyright 1973, distributed by Citadel Press, in *Out of My Later Years;* Princeton University Press for excerpts from *Einstein and Religion* by Max Jammer, copyright 1999; from *The New Quotable Einstein* by Alice Calaprice, copyright 2005; and *Albert Einstein: The Human Side,* selected and edited by Helen Dukas and Banesh Hoffman, copyright 1979 by the Albert Einstein Estate; Schocken Books for excerpts from *Einstein on Peace* edited by Otto Nathan and Heinz Norden, copyright 1968 by the Albert Einstein Estate; The Viking Press for excerpts from *Einstein* by Jeremy Bernstein, copyright 1973 by Jeremy Bernstein.

For material originally published prior to 1971, we have contacted the Albert Einstein Archives at the Hebrew University of Jerusalem to obtain permission to publish reproduction of material.

For unpublished material or for material originally published after 1971 or published in The Collected Papers of Albert Einstein, we have contacted Princeton University Press to obtain permission to publish reproduction of material.

Notes

No changes were made in the Einstein quotes in this book. All variant spellings of words remain as used in the original texts and translations. Much of the use of "man" may be due to mistranslations of the German *mensch,* meaning human.

COSMIC RELIGION
1 From Albert Einstein's letter to Milton M. Schayer, August 1927 (AEA 48-380).
2 Barnett, Lincoln, *The Universe and Dr. Einstein, 2nd Edition* (NewYork: William Morrow, 1974), 108.
3 Bernstein, Jeremy, *Einstein* (NewYork: Viking, 1973), 11. Also in: Nathan, Otto, Heinz Norden (eds.), *Einstein on Peace* (NewYork: Shocken, 1968), 282.
4 From Albert Einstein's letter to Norman Salit, March 4, 1950 (AEA 61-226).

BEGINNINGS
1 Einstein, Albert and Paul A. Schilpp (trans.), *Autobiographical Notes* (LaSalle and Chicago, IL: Open Court Publishing Company, 1979), 3, 5.

THE MEANING OF LIFE
1 Einstein, Albert, *The World As I See It* (NewYork: Wisdom Library of the Philosophical Library, 1949), 1.

SELF-PORTRAIT
1 First published in George S. Schreiber: *Portraits and Selfportraits* (Boston: Houghton Mifflin, 1936).

THE WORLD AS I SEE IT
1 Originally published in *Forum and Century* Vol. 84, 193–194, the thirteenth in the Forum series "Living Philosophies." Reprinted in *Living*

Philosophies (New York: Simon & Schuster, 1931), 3–7. Also in: Einstein, Albert Sonja Bargmann (trans.), *Ideas and Opinions* (New York: Crown, 1954), 8–11.

2 MacHale, Des, *Wisdom* (London: Prion, 2002).

RELIGION AND SCIENCE

1 Einstein, Albert "Religion and Science." *New York Times Magazine,* November 9, 1930, 1–4. Reprinted in: Einstein, Albert Sonja Bargmann (trans.), *Ideas and Opinions.* (New York: Crown, 1954), 36–40. Also in Einstein's book *The World as I See It* (New York: The Wisdom Library of the Philosophical Library, 1949), 24–28.

2 Moszkowski, Alexander and Henry L. Brose (trans.), *Conversations with Einstein* (New York: Horizon Press, 1970), 46.

3 French, Anthony P. (ed.), *Einstein: A Centenary Volume* (Cambridge, MA: Harvard UP, 1979), 66.

4 Barnett, Lincoln, *The Universe and Dr. Einstein* (New York: William Morrow, 1974), 108. Also in: Einstein, Albert, *Cosmic Religion, with other Opinions and Aphorisms* (New York: Covici-Friede, 1931).

5 Einstein, Albert, *Out of My Later Years* (Secaucus, NJ: Citadel, 1956), 26.

6 Vallentin, Antonia and Moura Budberg (trans.) , *The Drama of Albert Einstein* (Garden City, NY: Doubleday 1954), 259.

7 Einstein, Albert, *Cosmic Religion, with other Opinions and Aphorisms* (New York: Covici-Friede, 1931), 98.

8 Einstein, Albert, *Out of My Later Years* (Secaucus, NJ: Citadel, 1956), 29–30.

MORALITY AND VALUES

1 Dukas, Helen and Banesh Hoffman (eds.), *Albert Einstein: The Human Side.* (Princeton, NJ: Princeton UP, 1979), 69–70.

2 Ibid., 70–71.

3 French, Anthony P. (ed.), *Einstein: A Centenary Volume*. Harvard University Press 1979, 241

4 Einstein, Albert, *Mein Weltbild* (Amsterdam: Querido Verlag, 1934). Originally from a letter to an unidentified person, 1933.

5 Dukas, Helen and Banesh Hoffman (eds.) *Albert Einstein: The Human Side* (Princeton, NJ: Princeton UP, 1979), 95.

6 Einstein, Albert, *Out of My Later Years* (Secaucus, NJ: Citadel, 1956), 19.

7 Einstein, Albert and Sonja Bargmann (trans.), *Ideas and Opinions* (New York: Crown, 1954), 12. Also reprinted in: Einstein, Albert, *The World as I See It* (New York: The Wisdom Library of the Philosophical Library, 1949), 7–8

8 Dukas, Helen and Banesh Hoffman (eds.) Albert Einstein: *The Human Side* (Princeton, NJ: Princeton UP, 1979), 66.

MORAL DECAY

1 Einstein, Albert, *Out of My Later Years* (Secaucus, NJ: Citadel, 1956), 9–10. From a message to the Young Men's Christian Association, October 11, 1937.

2 MacHale, Des, *Wisdom* (London: Prion, 2002).

3 Ibid.

4 Ibid.

CHRISTIANITY AND JUDAISM

1 Einstein, Albert, *Out of My Later Years* (Secaucus, NJ: Citadel, 1956), 23.

2 Einstein, Albert, *Mein Weltbild* (Amsterdam: Querido Verlag, 1934). Also in: Einstein, Albert, *The World as I See It* (New York: The Wisdom Library of the Philosophical Library, 1949), 111–112. Statement for the Romanian Jewish journal *Renasterea Noastra*, January 1933.

3 Dukas, Helen and Banesh Hoffman (eds.) *Albert Einstein: The Human Side* (Princeton, NJ: Princeton UP, 1979), 96.

GOD

1 Hermanns, William, *Einstein and the Poet* (Brookline Village, MA: Branden, 1983), 132.

2 From a fall 1940 conversation recorded by Algernon Black. Einstein Archive 54–834. Also reprinted in: Calaprice, Alice, *The New Quotable Einstein* (Princeton, NJ: Princeton UP, 2005), 202

3 MacHale, Des, *Wisdom* (London: Prion, 2002).

4 Einstein, Albert, *Cosmic Religion, with other Opinions and Aphorisms* (New York: Covici-Friede, 1931), 102.

5 Vallentin, Antonia and Moura Budberg (trans.), *The Drama of Albert Einstein* (Garden City, NY: Doubleday, 1954), 102

6 Clark, Ronald W., *Einstein: The Life and Times* (New York and Cleveland, OH: The World Publishing Company, 1971), 19.

7 French, Anthony P. (ed.) *Einstein: A Centenary Volume* (Cambridge, MA: Harvard UP, 1979), 128.

8 Clark, Ronald W. *Einstein: The Life and Time* (New York and Cleveland, OH: The World Publishing Company, 1971), 19. Also in: French, Anthony P. (ed.) *Einstein: A Centenary Volume* (Cambridge, MA: Harvard UP, 1979), 67.

PRAYER

1 Dukas, Helen and Banesh Hoffman (eds.), *Albert Einstein: The Human Side* (Princeton, NJ: Princeton UP, 1979), 32–33.

MYSTICISM

1 Dukas, Helen and Banesh Hoffman (eds.) *Albert Einstein: The Human Side* (Princeton, NJ: Princeton UP, 1979), 40.

2 Ibid., 38

3 Ibid., 39

4 Ibid., 39

LIFE AND DEATH

1 Einstein, Albert and Sonja Bargmann (trans.), *Ideas and Opinions* (NewYork: Crown, 1954), 186.

THE ESSENCE OF RELIGION

1 Vallentin, Antonia and Moura Budberg (trans.), *The Drama of Albert Einstein* (Garden City, NY: Doubleday, 1954), 299.

THE INDIVIDUAL

1 Clark, Ronald W., *Einstein: The Life and Times* (NewYork and Cleveland, OH: The World Publishing Company, 1971), 622.

2 Einstein, Albert and Sonja Bargmann (trans.), *Ideas and Opinions* (NewYork: Crown, 1954), 43.

3 From a September 8, 1916 letter to Hedwig Born, wife of physicist Max Born. Einstein Archive 31–475. Reprinted in: Seelig, Carl (ed.) *Helle Zeit, Dunkle Zeit* (Zurich: Europa Verlag, 1956), 36. Also in: Calaprice, Alice, *The New Quotable Einstein* (Princeton, NJ: Princeton UP, 2005), 61.

4 Attributed to Einstein. Reprinted in: Calaprice, Alice, *The New Quotable Einstein* (Princeton, NJ: Princeton UP, 2005), 291.

5 From a June 1934 unpublished article on tolerance. Einstein Archive 28–280, and reprinted in: Calaprice, Alice *The New Quotable Einstein* (Princeton, NJ: Princeton UP, 2005), 266.

MORALS AND EMOTIONS

1 Einstein, Albert, *Out of My Later Years,* (Secaucus, NJ: Citadel, 1956), 15–20. Commencement address delivered at Swarthmore College, June 6, 1938.

CONSCIENCE

1 Vallentin, Antonia and Moura Budberg (trans.), *The Drama of Albert Einstein* (Garden City, NY: Doubleday, 1954), 300.

OF WEALTH

1 Statement for Viennese Weekly *Bunte Woche*, December 9, 1932. Reprinted as "Of Wealth" in *The World As I See It*.

2 From a sign hanging in Einstein's office at Princeton.

THE MENACE OF MASS DESTRUCTION

1 Einstein, Albert, *Out of My Later Years,* (Secaucus, NJ: Citadel, 1956), 204–206

2 From a December 14, 1930 speech to the New History Society in notes taken by Rosika Schwimmer. Reprinted as "Militant Pacifism" in: Einstein, Albert, *Cosmic Religion, with other Opinions and Aphorisms* (New York: Covici-Friede, 1931), 58.

3 Einstein Archive 48–479. Also in: Calaprice, Alice, *The New Quotable Einstein* (Princeton, NJ: Princeton UP, 2005), 158.

4 From an interview with Alfred Werner in *Liberal Judaism* 16, April - May 1949, 12. Einstein Archive 30–1104. Reprinted in: Calaprice, Alice, *The New Quotable Einstein* (Princeton, NJ: Princeton UP, 2005), 173 .

5 Einstein quoted in Konrad Bercovici, Pictoral Review, February 1933. Reprinted in: Clark, Ronald W., *Einstein: The Life and Times* (New York and Cleveland, OH: The World Publishing Company, 1971), 372–373.

6 May 23, 1946. Quoted in Nathan, Otto and Heinz Norden (eds.), *Einstein on Peace* (New York: Shocken,1968), 376. Reprinted in: Calaprice, Alice, *The New Quotable Einstein* (Princeton, NJ: Princeton UP, 2005), 175.

7 Attributed to Einstein.

8 Ibid.

9 Ibid.

10 From an address entitled "Science and Happiness" presented February 16, 1931, at the California Institute of Technology, Pasadena. Quoted in the *New York Times* February 17 and 22, 1931. Einstein Archive 36–320. Reprinted in: Calaprice, Alice, *The New Quotable Einstein* (Princeton, NJ: Princeton UP, 2005), 232–233.

11 Quoted on PBS television Nova documentary "Einstein", 1979.

WORLD PEACE

1 UN radio interview, 1950.

2 Einstein, Albert, *Cosmic Religion, with other Opinions and Aphorisms* (New York: Covici-Friede, 1931), 67.

SCIENCE AND RELIGION

1 Einstein, Albert, *Out of My Later Years* (Secaucus, NJ: Citadel, 1956), 21–30.

SCIENCE AND SOCIETY

1 Einstein, Albert, *Out of My Later Years* (Secaucus, NJ: Citadel,1956), 135–137.

ART AND CREATIVITY

1 1920. Quoted by Moszkowski, Alexander and Henry L. Brose (trans.), *Conversations with Einstein* (New York: Horizon, 1970), 184. Reprinted in: Calaprice, Alice, The New Quotable Einstein (Princeton, NJ: Princeton UP, 2005), 7–8.

2 November 15, 1950 regarding musician Ernst Bloch. Quoted in: Dukas, Helen and Banesh Hoffman (eds.) *Albert Einstein: The Human Side* (Princeton, NJ: Princeton UP, 1979), 77. Einstein Archive 34–332, and reprinted in: Calaprice, Alice, *The New Quotable Einstein* (Princeton, NJ: Princeton UP, 2005), 260.

3 Remark made in 1923. Recalled by Archibald Henderson *Durham Morning Herald* August 21, 1955. Einstein Archive 33–257. Reprinted in: Calaprice, Alice, *The New Quotable Einstein* (Princeton, NJ: Princeton UP, 2005), 230.

4 For a magazine on modern art, *Menschen. Zeitschrift neuer Kunst* 4, February 1921, 19. See also CPAE Vol. 7 Doc. 51, and reprinted in: Calaprice, Alice, *The New Quotable Einstein* (Princeton, NJ: Princeton UP, 2005), 252–253.

5 Einstein, Albert and Sonja Bargmann (trans.), *Ideas and Opinions* (New York: Crown, 1954), 32. From *Freedom, Its Meaning,* edited by Ruth Nan Anshen and James Gutmann (trans.) (New York: Harcourt, Brace and Company, 1940).

6 Attributed to Einstein.

IMAGINATION

1 "What Life means to Einstein," *Saturday Evening Post,* October 26, 1929. Also reprinted in: Calaprice, Alice, *The New Quotable Einstein* (Princeton, NJ: Princeton UP, 2005), 9.

CURIOSITY

1 Clark, Ronald W., *Einstein: The Life and Times* (New York: The World Publishing Company, 1971), 622.

2 MacHale, Des, *Wisdom* (London: Prion, 2002).

NATURE

1 To Margot Einstein in 1951, quoted by Hanna Loewy in A & E Television's Einstein Biography. VPI International, 1991. Also reprinted in: Calaprice, Alice, *The New Quotable Einstein* (Princeton, NJ: Princeton UP, 2005), 61.

2 Sullivan, Walter, "The Einstein Papers: A Man of Many Parts," New York Times March 29, 1972, 22 M.

ETERNAL MYSTERY

1 Einstein, Albert *Ideas and Opinions.* Sonja Bargmann (trans.). Crown Publishers, Inc., New York 1954, 292

THE GOAL OF HUMAN EXISTENCE

1 Einstein, Albert *Out of My Later Years.* The Citadel Press, Secaucus, New Jersey 1956, 260–261

MY CREDO

1 This article is an autumn 1932 speech to the German League of Human
Rights, Berlin. Also reprinted in the Appendix of: White, Michael and John
Gribbin, *Einstein, a Life in Science.* (New York: Dutton, 1994), 262-263.

THE RELIGIOUS SPIRIT OF SCIENCE

1 Einstein, Albert, *Mein Weltbild* (Amsterdam: Querido Verlag, 1934).
Reprinted in: Einstein, Albert and Sonja Bargmann (trans.), *Ideas and
Opinions* (New York: Crown, 1954), 40. Also reprinted in: Einstein, Albert,
The World as I See It (New York: Wisdom Library of the Philosophical
Library, 1949), 111–112.

Bibliography

Albrecht, Stephan, Reiner Braun and Thomas Held (eds.) *Einstein Weiterdenkin Thinking Beyond Einstein*. Bern, Switzerland: Peter Lang Publishing, 2006.

Babiak, Paul and Robert D. Hare. *Snakes in Suits*. New York: Regan Books, 2006.

Böll, Heinrich Stiftung (ed.) *Nuclear Power: Myth and Reality*. Sheffield: WISE, 2006.

Braun, Reiner and David Krieger (eds.) *Einstein—Peace Now!* New York: Wiley-VCH, 2005.

Brian, Denis. *Einstein, A Life*. New York: Wiley, 1995.

Caldicott, Helen. *Nuclear Power is not the Answer*. New York: New Press, 2007.

—. *The New Nuclear Danger*. New York: New Press, 2004.

—. *Nuclear Madness*. New York: WW Norton, 1994.

Chen, Joshua C. *Peace: 100 Ideas*. New York: CDA, 2003.

Chomsky, Noam. *What Uncle Sam Really Wants*. Boston: Odonian, 1992.

Deming, Barbara. *We Cannot Live without our Lives*. New York: Grossman, 1974.

Diamond, Louise Conari. *The Courage for Peace*. Berkeley, CA: Conari Press, 1999.

Dalai Lama and von Schoenborn, Felizitas. *Path of Wisdom, Path of Peace*. New York: Crossroads, 2005.

Eisler, Riane. *The Chalice and the Blade*. San Francisco: Harper One, 1988.

Gardiner, Robert W. *The Cool Arm of Destruction*. London: The Westminster Press, 1974.

Hartung, William D. *How Much are You Making on the War, Daddy?* New York: Nation Books, 2003.

Havel, Vaclav. *Living in Truth*. New York: Faber & Faber, 1990.

Horkheimer, Jack, O'Meara, Stephen, Steven, James, and Harrington, Rich (ilt.) *Stargazing with Jack Horkheimer.* Chicago: Cricket Books, 2007.

Isaacson, Walter. *Einstein, His Life and Universe.* New York: Simon & Schuster, 2007.

Krieger, David and Ikeda, Daisaku. *Choose Hope.* Santa Monica, CA: Middleway Press, 2002.

Lifton, Robert J. and Falk, Richard. *Indefensible Weapons.* New York: Basic Books, 1982.

Looking Horse, Chief Arvol. *White Buffalo Teachings.* Williamsburg, MA: Dreamkeepers, 2001.

Mahesh, Maharishi. *Maharishi's Absolute Theory of Defense Yogi.* Fairfield, IA: Maharishi University of Management Press, 1996.

Miller, Arthur I. *Einstein, Picasso.* New York: Basic Books, 2002.

Mitchell, Edgar. *The Way of the Explorer.* New York: Putnam, 1996.

Mortenson, Greg and Relin, David Oliver. *Three Cups of Tea.* New York: Penguin, 2007.

Nhat Hanh, Thich. *Being Peace.* Berkeley, CA: Parallax Press, 2005

Oates, Robert M. *Permanent Peace.* Fairfield, IA: Maharishi University of Management Press, 2002.

Parry, Danaan. *Warriors of the Heart.* California: Earthstewards Network, 1997.

Rapoport, Anatol. *The Origins of Violence.* London: Paragon, 1989.

Rowe, David E. and Schulmann, Robert (eds.). *Einstein on Politics.* Princeton, NJ: Princeton UP, 2007.

Saint-Exupery, Antoine de. *The Little Prince.* New York: Harcourt, 1968.

Satir, Virginia. *New Peoplemaking.* Palo Alto, CA: Science and Behavior Books, 1988.

Schilp, Paul A. (ed.) *Albert Einstein: Philosopher–Scientist.* New York: MJF Books, 2001.

Stout, Martha. *The Sociopath Next Door.* NewYork: Broadway, 2005.

Tuttle, Will. *The World Peace Diet.* Brooklyn, NY: Lantern, 2005.

Trumbo, Dalton. *Johnny Got his Gun.* NewYork: Bantam, 1984.

Tyson, Neil DeGrasse. *E = Einstein.* NewYork: Sterling, 2007.

Wallace, Keith and Marcus, Jay B. *Victory Before War.* Fairfield, IA: Maharishi University of Management Press, 2005.

Yoder, John H. *The Politics of Jesus.* Grand Rapids, MI: Eerdmans Publishing, 1994.

Zajonc, Arthur (ed.) *New Physics and Cosmology.* Oxford: OUP, 2004.

A Short Selected List of Einstein's Writings

Einstein, Albert. *Cosmic Religion, with other Opinions and Aphorisms.* NewYork: Covici/Friede, 1931.

___. *Essays on Science.* NewYork: Philosophical Library, 1934.

___. *The World As I See It.* London: John Lane, 1935.

___. *The Meaning of Relativity.* NewYork: Crown, 1950.

___. R. Fürth (ed.) *Investigations on the Theory of the Brownian Movement.* Mineola, NY: Dover, 1956.

___. *Out of My Later Years.* NewYork: Citadel, 1956.

___. *The Principle of Relativity.* Dover Publications 1956.

___. L. Infeld. *The Evolution of Physics.* Cambridge UP, 1961.

___. *Relativity.* NewYork: Crown, 1961.

___. *Ideas and Opinions.* NewYork: Dell Publishing, 1973.

___. Robert Schulmann (contrib.), Alice Calaprice (ed.), Evelyn Einstein (intr.). *Dear Professor Einstein.* Amherst, NY: Prometheus, 2002.

___. Alice Calaprice (ed.), Freeman Dyson (forew.) *The Ultimate Quotable Einstein.* Princeton, NJ: Princeton UP, 2011.

Image Credits

ADDITIONAL CAPTIONS

Page viii–ix **POLAR AURORAE,** also known as the aurora borealis (or the northern lights) or the aurora australis (or the southern lights) are one of nature's most spectacular and colorful light shows. Depicted here in duotone, this shot called "Angel Fire" (Nordly's winner of the Year Photo Contest 2005) retains much of its beauty.

Page xxiv–1 **GREAT ANDROMEDA GALAXY,** M 31 is the nearest major galaxy to our own Milky Way. Our galaxy was thought to look much like Andromeda, though recent studies indicate it is a barred spiral. Together these two dominate the local group of galaxies. The diffuse light from Andromeda is caused by hundreds of billions of stars that compose it. The several distinct stars that surround Andromeda's image are actually stars in our galaxy that are well in front of the background object. This beautiful picture is a high resolution mosaic of 40 separate grayscale frames taken from September to November 2002. The telescope used was a 12.5" Ritchey Chretien Cassegrain. The image was selected as one of the greatest astronomical images of the last thirty years by *Astronomy Magazine.*

Page 18–19 **HELIX NEBULA, NGC 7293.** This celestial object is the nearest planetary nebula to the Sun. The different colors are natural and reveal various ionization levels within the shell of matter ejected from a central star. Excited (heated) oxygen atoms give the middle region its bluish hue, while the outer red is predominately nitrogen and hydrogen. The smallest radial blobs inside the red shell are about 150 astronomical units across (150 times the Earth-Sun distance). They give the lovely object its alternative name, Sunflower nebula.

Page 50–51 **M 42, GREAT ORION NEBULA.** An amateur astronomer took this great shot at Joshua Tree national park in November 2000.

Page 64–65 **THE MICE.** Located in the Coma cluster of galaxies, about 300 light-years away, this interacting pair of galaxies is known as the Mice (NGC 4676), because of their long tails of stars and gas. They will eventually merge into a single galaxy.